AI is Coming for Your Business

AI is Coming for Your Business

How Smart Owners Are Using AI to Cut Costs, Win Customers, and Outperform the Competition

First Edition published 2026
Published in the United States by Schiano Products & Services LLC., Cumming, GA 30040

SP&S LLC.
Business Communication Services

- Digital Media
- Contact Center
- Leadership
- Customer Engagement
- Revenue Growth

For more information, email **mike@spscontact.com**

www.spscontact.com

ISBN: 979-8-90417-206-0

To my four beautiful children who inspire me to be better every day. You are inheriting a world that needs fixing and it will require every ounce of your creativity, knowledge, ingenuity, inspiration, and tenacity. I am grateful that you all have an abundance of those attributes and you are all truly good people.

Foreword

"It no longer requires a multi-million-dollar budget to get AI going in your company. It represents an opportunity to level the playing field for smaller companies."
— Nichole Jordan, Managing Partner, Grant Thornton LLP

"AI is not only for engineers. It brings changes in the dynamic of business, and we have to adapt or die."
— Satya Nadella, CEO, Microsoft

"Start using AI deeply. It is a huge intelligence amplifier."
— Reid Hoffman, Co-founder, LinkedIn

"AI is a catalyst for innovation, pushing the boundaries of what's possible. Leaders who embrace AI into their company culture avoid digital irrelevance."
— Contentstack leadership commentary

"It is difficult to think of a major industry that AI will not transform."
— Andrew Ng, Co-founder, Google Brain

"The true power of AI lies not in replacing humans, but in working alongside us to achieve what neither can do alone."
— Sebastian Thrun, founder, Udacity and Kitty Hawk

"AI can make us faster, smarter and more informed, but it can't make us more caring, empathetic or trustworthy. That's still our job."
— Shep Hyken, customer service expert and author

"The playing field is poised to become a lot more competitive, and businesses that don't deploy AI and data to help them innovate in everything they do will be at a disadvantage."
— Paul Daugherty, Chief Technology & Innovation Officer, Accenture

Table of Contents

Introduction

AI Is Coming for Your Business: The Wake-Up Call You Can't Afford to Ignore

If you own or run a business, you can feel it.

Something is shifting.

Leads are harder to close.

Customers expect everything faster.

Your inbox never stops.

Your team is maxed out.

You might blame the economy, competition, or just think it's how things are now.

But the real shift is deeper and more permanent:

You are now competing against AI-powered businesses.

This isn't a political shift.

It isn't a cultural movement.

It isn't a trend on social media.

It's technological.

Artificial intelligence isn't "the next big thing."

It's the biggest transformation of your working lifetime—bigger than the internet, smartphones, or social media.

And this time, small and medium-sized businesses are the ones directly in the blast radius.

This book exists for one reason:

To make sure your business doesn't get left behind—and to show you how to turn AI into your unfair advantage.

Why This Matters Right Now

Ten years ago, AI was a research project.

Five years ago, it was a toy for big tech companies.

Today, it's a powerful tool for small businesses.

Right now, your competitors—local, national, and online—are quietly plugging AI into:

- customer service
- marketing and content
- sales and follow-up
- scheduling and reminders
- invoicing and payments
- hiring and screening
- financial forecasting
- email and inbox management
- documentation and admin work

They're not bragging about it. They're just getting:

- 50–70% time savings on routine work
- instant responses to leads and customers
- higher conversion rates
- fewer mistakes and callbacks
- a stronger online presence
- lower labor and admin costs
- better, more consistent customer experiences

These are not "consultant numbers."

They're real improvements from real businesses.

The businesses using AI today are quietly pulling away.

The ones ignoring it are quietly falling behind.

Most won't realize what happened until it's too late.

The Harsh Truth: You're Competing Against Augmented Businesses

If you run:

- A home service company
- a professional practice or freelance consultancy
- a restaurant or cafe
- a medical or therapy office
- a salon or spa
- a real estate team or brokerage
- a construction or trades business
- a retail store or boutique
- a gym or fitness studio
- an accounting or Tax Practice

AI is already affecting your industry.

Not "someday."

Not "when the tech catches up."

Now.

Competitors who adopt even basic, inexpensive AI tools are outpacing those who don't in every meaningful way:

- faster replies
- tighter operations
- better documentation
- smarter marketing
- more consistent follow-up
- smoother customer experiences

If you're still running purely on human time and human memory, you are trying to outrun a car on foot.

AI Isn't Going to Replace You—But an Owner Using AI Will

Here's the core truth of this book:

AI will not replace small businesses.

Business owners who use AI will replace those who don't.

In a world where your customers expect:

- instant answers
- rapid delivery
- flawless service
- consistent communication
- fresh, relevant content
- accurate information

- personalized experiences

...a business trying to do all of that manually simply cannot keep up.

The scary part is that your competitors don't need to be smarter than you.

They just need to be augmented.

- They'll respond faster.
- They'll follow up more consistently.
- They'll make fewer mistakes.
- They'll run leaner and more profitably.
- They'll deliver better customer service.
- They'll make decisions based on data, not guesswork.
- They'll build systems instead of constantly putting out fires.

All without hiring a big team.

That's the real competitive threat.

The Real Reason Small Business Owners Fear AI

Over the last several years, I've talked with thousands of business owners—on calls, in workshops, in mastermind groups, and over coffee.

When we strip away the buzzwords, their fears fall into the same five buckets:

1. "I have no idea where to start."
2. "This is going to be too expensive."

3. “I’m afraid AI will break something or create legal/compliance headaches.”
4. “I’m worried AI will replace my employees—or me.”
5. “I don’t think I’m tech-savvy enough to learn this.”

Those are not technical problems.

They’re emotional ones.

And they’re valid.

You’ve spent years building your business.

You can’t afford to gamble it on unproven tools, shady vendors, or half-understood buzzwords.

So let me be clear:

- This book is not a love letter to AI.
- It’s a field manual for using AI in a way that is safe, sane, practical, and profitable for real-world businesses.

We’ll address every one of those fears head-on.

Because once owners see how AI works in a small business without the hype or horror stories, something clicks:

They realize AI isn’t a threat. It’s the biggest operational upgrade they’ve seen in decades.

AI Levels the Playing Field Like Nothing Before

For the first time in history:

- A small business can market like a big brand.
- A solo operator can run like a five-person team.
- A local shop can provide world-class customer service.
- A startup can build workflows that used to require enterprise software and consultants.
- A brick-and-mortar business can automate back-office work that once required an office full of staff.

AI is a force multiplier for the underdog.

AI doesn't care about your headcount, your location, or your square footage.

In the right hands, it lets the little guy not just compete.

—but often win.

The Goal of This Book: AI Without the Overwhelm

This book is not written for engineers.

It's not written for Silicon Valley.

It's not written for people who enjoy reading technical documentation for fun.

It's written for you if you:

- run lean, with a small or overworked team
- wear ten hats: sales, marketing, operations, HR, finance, customer service
- work more hours than you'll admit
- never feel fully caught up
- know you can't ignore AI—but feel overwhelmed, skeptical, or both

You don't need another textbook about algorithms.

You need a playbook you can actually use.

So this book gives you what few other books provide:

- simple, concrete steps—not theory
- practical tools you can start with today
- real examples from real businesses
- plug-and-play workflows you can copy and adapt
- safe, beginner-friendly ways to test AI in your business
- a clear roadmap from "AI curious" to "AI-powered."
- Usable AI prompts to use right away in your business to 10x your success.

If you can send an email, you can use AI.

If you can copy and paste, you can use AI.

If you can run your business today, you can absolutely run it better with AI tomorrow.

What You'll Learn (and Actually Use)

Here's what we're going to cover together:

- The biggest myths keeping small businesses from using AI—and the truth behind them
- How to quickly spot your highest-ROI AI opportunities
- The first workflow you should automate (and why)

- The core tools every small business should consider (without overspending or overbuilding)
- How AI can improve your marketing, sales, customer service, operations, HR, and finance
- How to train your team so they want to use AI instead of fearing it.
- How to create an AI culture that empowers people instead of threatening them
- How to avoid scams, bad vendors, and unsafe tools
- How to write a simple AI policy that protects your business
- How to build a realistic, practical 12-month AI transformation plan for your company

By the final chapter, you will:

- understand where AI fits into your business
- know exactly what to do first, second, and third
- have workflows you can start implementing the same week
- feel confident talking about AI with your team, vendors, and peers
- be ahead of at least 90% of your competitors who are still "thinking about it."

You're Not Behind—Yet.

One of the most common questions I hear is:

"Am I already too late?"

The honest answer?

No.

But the window is closing.

Right now, AI is still in the “early adopter advantage” phase.

Those who start now will:

- adapt faster
- win more customers
- strengthen margins
- build more efficient teams
- be ready for what’s coming next

Those who wait will spend the next decade trying to catch up to competitors who quietly got serious about AI while they were still debating it.

You’re not behind if you start now.

Your Business Has Two Futures

Five years from today, when someone tells the story of your business, one of two realities will be true.

Story 1:

You fell behind because you resisted change, stayed overwhelmed, and kept doing everything the old way until the gap became too big to close.

Story 2:

You modernized, streamlined, and grew.

You used AI to give your team superpowers.

You became the business in your space that others quietly studied and tried to copy.

This book exists to push you toward Story 2.

What Comes Next

Here's how we're going to do it.

In the chapters ahead, we will:

- cut through the hype and fear around AI
- dismantle the myths that keep most owners frozen
- help you identify what to automate first (and what to leave alone—for now)
- walk you step-by-step through your first 30 days with AI
- show you real-world tools that are affordable and actually work
- show you how to use AI in marketing, sales, customer service, operations, HR, and finance
- help you build an AI-enabled workforce that feels supported, not threatened
- help you avoid the biggest mistakes and risks
- guide you through building a complete 12-month AI roadmap for your business

By the end, you'll have more than information.

You'll have a plan.

Your business is about to become:

- faster
- smarter
- more profitable
- more efficient
- more modern
- more competitive

This is your moment.

Turn the page.

Let's build your AI-powered future—one smart step at a time.

A Quick Note on Tools, Examples, and Change

Throughout this book, you'll see references to specific AI tools, platforms, and services. These examples are included to illustrate what's possible and to make implementation easier—not because any single tool is required or endorsed above all others.

AI technology evolves rapidly. Tools will change. Features will improve. New platforms will emerge.

The principles, workflows, and strategies in this book are designed to outlast any specific software. Focus on understanding how to think about AI inside your business, not just which tools you use today.

This book is for educational purposes and practical business guidance. You are always responsible for how you apply these ideas within your own operations, industry, and regulatory environment.

Now, let's get to work.

A Moment Before We Go Further

Before we move on, I want you to pause for a moment.

Not because you're behind.
Not because you need to "catch up."
And definitely not because you need to understand everything right now.

But because if you're like most business owners I talk to, your head is already spinning a little.

You're probably thinking about:

- the fires waiting for you when you put this book down
- the emails you haven't answered yet
- the customers who expect faster responses
- the team that's already stretched thin
- the nagging feeling that you're doing *everything*—and still falling behind
  That feeling matters.

Because the goal of this book is not to turn you into a tech expert. And it's not to push you into reckless change.

The goal is simpler and more important.

It's to help you **regain control**.

Control of your time.
Control of your operations.
Control of how your business runs day to day.

AI is not here to make you work faster just so you can work more. Used correctly, it does the opposite.

It removes friction. It clears mental space. It takes weight off your shoulders so you can focus on what actually matters.

Before we dive into strategies, tools, and workflows, I want you to reflect on just three questions:

- Where does your business feel heaviest right now?
- What work drains your energy without truly moving the needle?
- If you had five extra hours a week, where would you want them to go?

You don't need perfect answers. You don't need a plan yet. You don't need to solve anything in this moment.

Just notice.

Because the owners who get the most out of AI aren't the ones chasing every new tool. They're the ones who are clear about what *shouldn't* require so much effort in the first place.As we move forward, keep this in mind:

You don't need to transform your entire business. You don't need to automate everything. You don't need to change who you are or how you lead.

You only need to start removing the work that never should have consumed you to begin with.

That's what this book is really all about.

Now, let's talk about the divide that's already reshaping small business and how to make sure you're on the right side of it.

Turn the page.

Chapter 1: The Great Business Divide: AI Adopters vs. AI Holdouts

Why the Next Five Years Will Determine Which Businesses Survive and Which Disappear

There are moments in history when the landscape of business shifts so dramatically that the old rules simply stop working. We've lived through a few of them: electricity, automobiles, the telephone, computers, the internet, smartphones, social media. Each one created winners and losers — not based on intelligence, talent, or luck, but on one thing:

Who adapted first!

AI is now creating the next division.
Not between big companies and small ones.
Not between startups and legacy brands.

But between **business owners who adopt AI** and **business owners who hesitate.**

The scary part?
The divide is widening every single month.

This chapter is your wake-up call — the one most business owners won't get until it's too late.

INTRODUCTION

The New Divide No One Wants to Admit Is Real

A new line has been drawn in the world of small business quietly, quickly, and without permission.

On one side are the owners who've embraced AI: the Adopters. They respond faster, operate leaner, deliver better experiences, and do in hours what used to take days.

On the other side are the Holdouts: the owners who want to wait, watch, or "see where things go."

They don't realize they're falling behind until the numbers stop working, the costs pile up, and customers start disappearing.

For decades, the formula for small business success was familiar: location, price, reputation, service.

That formula is now obsolete.

Today, **speed + automation + intelligence = survival.**

The only real question is:

Which side of the divide will your business stand on?

Because the data is uncomfortably clear:

- Employees using AI complete tasks nearly **twice as fast** with **higher accuracy**
- Businesses using AI cut costs by **10–40%**
- Customer satisfaction increases **25–60%**

- Lead conversion improves **20–50%**
- Admin workload drops **30–70%**

These aren't futuristic predictions.
These are current, real-world results.

And they are reshaping the small business landscape faster than any previous technology.

This chapter isn't here to scare you — it's here to show you what's happening and give you the advantage.

By reading it, you're already doing what most competitors won't:

You're preparing.

SECTION 1 — Why AI Is the New Electricity

If this moment feels eerily familiar, it's because we've been here before.

Imagine a business owner in 1920 saying:

- "Do we really need electricity?"
- "Gas lighting works fine."
- "Electricity is too new — what if it breaks something?"
- "We don't have the time or money to rewire."

It sounds ridiculous now.
But those conversations were real.

And that's exactly where many business owners are with AI in 2025.

Electricity didn't just replace gas lamps.
It reshaped everything:

- how factories ran
- how customers were served
- how products were made
- how fast businesses could scale
- how efficiently teams could work

AI is doing the same — but at **10× the speed.**

Here's the simplest way to understand it:

- **Electricity multiplied human strength.**
- **AI multiplies human intelligence.**
- Electricity automated physical labor.
- AI automates mental labor.
- Electricity lowered operating costs.
- AI lowers *every* category of business cost.
- Electricity made businesses scalable.
- AI makes **small teams unstoppable.**

And just like electricity, AI has moved through its three historical phases:

1. **Luxury** — only large companies used it
2. **Advantage** — early adopters began pulling ahead
3. **Infrastructure** — every business was forced to get on board

AI is now entering Phase 3.

It is becoming the new baseline for survival.

SECTION 2 — The Existential Threat Facing SMBs

Let's sharpen this even further:

AI isn't going to replace your business.
AI-powered competitors will.

The threat isn't robots.
The threat is a competitor across town who now operates like a 20-person team thanks to AI.

Here's how the divide plays out in the real world:

1. Speed

AI-enabled companies respond in **seconds.**
Holdouts respond in **hours or days.**

Customers don't wait.
They choose the fastest reply.

2. Productivity

AI-assisted employees do **3–10× more work** without burning out.
Holdout employees drown in repetitive tasks.

3. Cost Efficiency

AI cuts expenses in:

- marketing
- admin
- scheduling
- customer service
- bookkeeping
- hiring
- content creation

Holdouts maintain bloated processes and higher payroll costs.

4. Customer Experience

AI-enabled businesses deliver:

- instant answers
- consistent communication
- accurate information
- no dropped balls

Holdouts make mistakes, get overwhelmed, and frustrate customers without meaning to.

5. Innovation and Agility

AI businesses improve weekly.
Holdouts improve only when they "have time" — which they never do.

This is why the divide is accelerating.
Every month an AI Adopter gets faster, leaner, and more profitable...

...while Holdouts fall further behind.

Real-World Impact (That You Can't Ignore)

Hard data from Harvard, MIT, BCG, McKinsey, and Accenture all show the same pattern:

Productivity Lift: 38–66%

People complete tasks faster and better with AI.

Lead Conversion Boost: 20–50%

AI fixes the #1 sales killer: slow follow-up.

Cost Reduction: 10–40%

AI eliminates repetitive administrative labor.

Customer Satisfaction Increases: 25–60%

AI-enabled response times and accuracy dramatically improve service.

These numbers decide who stays in business.

Not in theory but in practice.

SECTION 3 — Case Study:

How a 7-Person Cleaning Company Grew 30% with AI

Meet Summit Home Cleaning, a small cleaning company in the Midwest.
They didn't have investors.
They didn't have tech experts.

They had seven employees, one overwhelmed office manager, and a frustrated owner.

They also had problems you might recognize:

- chaotic scheduling
- missed calls
- inconsistent follow-up
- slow hiring
- uneven marketing
- constant admin pressure

In late 2023, they started testing AI cautiously.

Six months later:

1. Appointments increased 22%

AI handled:

- instant responses
- quoting
- scheduling
- FAQs
- rescheduling

2. Repeat customers increased 30%

AI reminders created consistent retention.

3. Scheduling mistakes went to zero

AI drafted confirmations, spotted conflicts, and flagged errors.

4. Admin hours dropped by 40%

Paperwork practically disappeared.

5. Hiring became easier

AI screened applicants and drafted job descriptions.

The owner summed it up perfectly:

"AI didn't replace my team. I'd say it likely saved them, and me."

Summit now runs:

- faster
- leaner
- more efficiently
- more profitably

And they're crushing competitors with twice their staff.

SECTION 4 — Case Study:

The Photography Business That Closed Because Competitors Adopted AI First

Golden Hour Photography was a respected, 12-year-old photography business with loyal clients and solid reviews.

But the industry changed, and they didn't.

Competitors began using AI to:

- respond instantly

- follow up consistently
- edit faster
- deliver galleries within days
- automate bookings
- create AI-enhanced sample albums
- send reminders automatically

Golden Hour kept doing things the old way:

- manual editing
- delayed email replies
- handwritten contracts
- no follow-up system
- inconsistent delivery times
- no automation
- outdated website

Within 18 months:

- bookings dropped 50%
- revenue fell 42%
- admin consumed the owner
- customer complaints increased
- competitors surged ahead

Golden Hour didn't lose to AI.
They lost to **photographers using AI.**

They closed quietly in early 2025.

It wasn't a lack of talent.
It was a lack of adaptation.

This is the divide in real life.

A Moment for Reflection

These are examples of the outcomes that are possible for businesses of all sizes and various industries. As the owner of a business with employees, even companies of one person, you, the owner, carry the responsibility to look up and see the pain points impacting your business' growth and seek assistance, human or technical.

It is a very real responsibility that many owners miss.

Too many business owners facing problems get bogged down in the work thinking keeping busy is the answer.

Afterall, hard work is what got you to where you are today so, working more hours, later nights, less sleep, and pushing harder would seem like a very natural response to problems. You are doing what you know to do in the face of challenge.

You should be commended for doing something.

But, as you know, action can't be confused with progress. Keeping busy is not the same as taking effective action to accomplish objectives.

Let me tell you about a moment that changed how I run businesses completely.

It didn't happen in a boardroom.
It didn't come from a book.
It didn't come from some breakthrough technology.

It came from a problem that should have been simple but wasn't.

I was deep in the weeds like most business owners find themselves at some point. We had systems layered on top of systems. Reports feeding other reports. People chasing tasks that shouldn't have existed in the first place. Everyone was working hard. Really hard.

And yet... things weren't getting better.

Response times were slow.
Leads were slipping.
The team felt stretched.
And I kept telling myself the same lie most business leaders tell:

"We just need to push a little harder."

So that's what we did.

More hours.
More meetings.
More tracking.
More complexity.

And somehow there were worse results.

I remember sitting there late one night, looking at a dashboard that had more data than I could realistically act on. It looked impressive. It felt sophisticated.

But

...it was useless.

That was the moment it hit me.

Not as some grand realization. It was a quiet, uncomfortable truth:

We had built something complicated instead of something effective.

All that sophistication with the dashboards, the workflows, the layers of oversight wasn't helping us move faster. It was slowing us down. It was giving us the *feeling* of control without actually creating results.

So, I did something that, at the time, felt almost reckless.

I started stripping things away.

Not optimizing.
Not tweaking.
Not adding another tool.

Removing!

We cut reports no one used.
We simplified communication.
We reduced steps in workflows.
We stopped tracking things that didn't directly impact revenue or customer experience.

And then we focused on one thing:

What is the simplest action that actually moved the business forward?

That question became a filter.

If something didn't clearly drive:

- faster response
- better customer experience
- more revenue
- or less wasted time

...it didn't belong.

What happened next surprised me.

Within weeks things started to change.

Response times improved.
The team felt lighter, not heavier.
Decisions became faster.
And most importantly, results started to show up again.

Not because we worked harder.

Because we stopped making things harder than they needed to be.

That's when I learned something that has stayed with me ever since:

Complexity is often a sign you've lost control.
Simplicity is how you get it back.

And here's where this becomes incredibly important as we talk about AI.

Most business owners assume AI is going to make things more complicated.

More tools.
More systems.
More things to learn.

But the truth is the opposite if you use it correctly.

AI, at its best, is not about adding layers.

It's about removing friction.

It's about taking something that currently takes:

- five steps
- three people
- and two hours

...and turning it into:

- one step

- one person
- and five minutes

That's not sophistication.

That's simplicity at scale.

And the businesses that win in this next phase won't be the ones with the most advanced systems.

They'll be the ones that ask a better question:

"How do we make this simpler?"

Because once you understand that...

You stop chasing tools.
You stop chasing trends.
You stop trying to "do AI."

And instead, you start building a business that is:

- faster
- clearer
- easier to run
- and far more profitable

That shift from complexity to simplicity is where everything begins.

And it's exactly where we're going next.

SECTION 5 — The Speed Gap: The Silent Killer of SMBs

AI doesn't just speed up tasks.
It speeds up **everything**.

- Decision-making
- Communication
- Operations
- Sales
- Marketing
- Hiring
- Training
- Customer service
- Problem-solving
- Planning
- Documentation

When one business compounds these gains monthly, and another doesn't, the distance between them grows exponentially.

AI doesn't just make a business faster.

It makes a business **impossible to compete against** using traditional methods.

SECTION 6 — Why Some SMBs Still Avoid AI (And Why They Shouldn't)

Most business owners who resist AI aren't stubborn.

They’re overwhelmed.

Their fears are legitimate:

- “What if I break something?”
- “What if this messes up my customers?”
- “What if my team pushes back?”
- “What if it’s too expensive?”
- “What if it replaces my job?”

These fears come from **uncertainty**, not stupidity.

And here’s the truth:

When implemented correctly, AI:

- makes your staff more valuable
- reduces workloads
- improves customer satisfaction
- lowers costs
- reduces errors
- improves consistency
- gives owners their time back
- grows the business

AI isn’t a threat.
AI is an upgrade.

A big one.

SECTION 7 — The Business Advantage Graph (Explained Visually)

Picture a line graph with two lines starting at the same point:

Business A — The AI Adopter
The line curves upward every month:

- compounding savings
- compounding efficiency
- compounding marketing output
- compounding customer loyalty
- compounding speed

Business B — The Holdout
The line stays flat, then dips:

- slower responses
- missed opportunities
- overwhelmed staff
- inconsistent operations
- rising costs

After 12 months, the gap is noticeable.
After 24 months, it's significant.
After 36 months, it's permanent.

AI creates **competitive separation** — and once the gap is wide, it rarely closes.

SECTION 8 — Why This Book Begins Here

This book opens with the divide because this is the truth most business owners never hear clearly:

AI isn't:

- a tech fad
- a trend
- a Silicon Valley toy
- a "nice-to-have"
- something to revisit later

AI is now:

- infrastructure
- the new workforce multiplier
- the new customer experience engine
- the new operations backbone
- the new marketing engine
- the new administrative force
- the new hiring assistant
- the new revenue accelerator

You don't need to become a tech expert to survive this shift.

But you do need to avoid becoming an AI Holdout.

The business graveyard is already filling with companies that waited too long.

MOVING FORWARD

The Divide Is Real — But Your Future Is Still in Your Hands

This chapter wasn't written to intimidate you.
It was written to prepare you.

Because the great divide is already here:

- **AI Adopters** who thrive, grow, and scale
- **AI Holdouts** who fall behind and eventually disappear

But unlike previous technological revolutions, **you still have a choice.**

By reading this book — and especially this chapter — you've already taken the first step.

You've chosen learning over fear.
You've chosen leadership over hesitation.
You've chosen to build a business that can win in the next decade.

Now, it's time to make AI simple, clear, and manageable.

Turn the page.

Let's remove the fog, break down the myths, and show you exactly what AI is — and how to use it to build a business that's faster, smarter, and built for the future.

CHAPTER 2: Clearing the Fog: What AI Actually Is (and Isn't)

A Simple, Practical Explanation for Business Owners Who Don't Want Tech Jargon. They Want Clarity.

INTRODUCTION

AI Confusion Is the #1 Barrier Keeping Small Businesses Stuck

Ask most business owners what they think about AI and you'll see the same mix on their faces:

- curiosity
- fear
- confusion
- uncertainty
- overwhelm

The problem isn't that AI is too complex.
The problem is that it's been **explained horribly**.

Most of what you hear about AI comes from:

- "Tech bros" chasing hype
- overcooked marketing claims
- confusing headlines
- jargon-heavy articles

- fear-based news segments
- YouTube “gurus” promising shortcuts
- Silicon Valley buzzwords that mean nothing in real life

None of that helps a small business owner who just wants to:

- serve customers
- grow revenue
- reduce stress
- run a tighter operation
- stop drowning in work

You don’t need another hand-wavy explanation of “algorithms” and “neural networks.”

You need clarity.
You need plain language.
You need to know what AI can do for your business **today**.

That’s what this chapter gives you.

SECTION 1 — What AI Actually Is: The Simplest Definition You’ll Ever Read

Let’s throw out the noise and start fresh.

Here’s the simplest, most useful definition of AI for a business owner:

AI is software that can understand information, recognize patterns, and generate useful responses or actions — without needing you to spell out every step.

It's not:

- robots taking over
- sci-fi magic
- a black box that only coders can use
- an instant job killer

Instead, think of AI as a **powerful assistant** that can:

- read
- write
- summarize
- organize
- analyze
- respond

...with incredible speed and consistency.

If machines automated **physical** labor, AI automates **mental** labor:

- drafting
- planning
- reviewing
- sorting
- scheduling
- answering
- summarizing

Once you see AI as "automation for thinking tasks," everything else becomes easier to understand.

SECTION 2 — AI vs Automation vs Machine Learning (Without the Jargon)

A lot of confusion comes from three terms being thrown together:

- automation
- Machine Learning
- AI

They're related, but not the same.

Here's the business-friendly breakdown.

1. Automation

Automation = software following rules you set.

Examples:

- "If a customer fills out this form → send this email."
- "If payment fails → send a reminder."
- "If an invoice is past due → notify accounting."

Automation is:

- predictable
- rule-based
- rigid

It does exactly what you tell it to do, nothing more.

2. Machine Learning (ML)

Machine learning = software that improves over time by spotting patterns in data.

Examples:

- Google Maps predicting traffic
- Netflix recommending shows
- Credit card companies flagging suspicious charges

ML is about:

- pattern detection
- predictions
- continuous improvement

But it generally doesn't talk to you.

3. Modern AI (LLMs and Generative AI)

AI = software that can understand language, generate content, and make decisions in flexible, human-like ways.

Examples:

- ChatGPT writing emails, scripts, or blog posts
- Claude summarizing long documents
- AI tools analyzing financial data and trends
- AI agents replying to customers, qualifying leads, or booking appointments

AI systems don't just follow rules — they **adapt.**

They use context, language, and prior examples to respond in a way that feels more like a knowledgeable assistant than a computer program.

The Business-Owner Summary

If you remember nothing else from this section, remember this:

- **Automation** = "Do this when X happens."
- **Machine Learning** = "Learn from past examples."
- **AI** = "Think and respond like a smart assistant."

Once you separate these three in your mind, the whole AI conversation gets less intimidating.

SECTION 3 — Why AI Keeps Getting Better (While Costs Keep Dropping)

One reason AI feels overwhelming is because every few months, it seems like there's a "new version" that's faster, smarter, and cheaper.

That's not your imagination.
AI really is improving at a ridiculous pace.

There are three big reasons why.

1. Every User Interaction Helps Improve the Model

At a high level, every meaningful interaction with AI helps developers improve it.

Your private business data can be protected and kept separate (and should be — we'll talk about that in the safety chapter). But across billions of interactions, AI systems get better at:

- understanding language
- following instructions
- avoiding errors
- providing relevant outputs

The more people use AI, the smarter and more useful it gets.

2. Computing Power Keeps Getting Cheaper and Better

AI runs on specialized computer chips.

Those chips keep getting:

- faster
- more efficient
- cheaper

Which means:

- more power for less money
- more features without massive cost
- better performance on everyday devices

The kind of AI that used to require a research lab now runs in your browser for a monthly subscription.

3. Competition Is Forcing Rapid Innovation

The biggest tech companies in the world — OpenAI, Anthropic, Google, Meta, and others — are in an arms race.

That race drives:

- constant feature updates
- better reasoning
- improved memory and context handling
- more integrations with tools you already use
- lower prices over time

Looking at the next few years, you can assume:

- AI will be significantly **cheaper**
- AI will be significantly **smarter**
- AI will be **built into** almost every tool (email, CRM, docs, scheduling, accounting, etc.)

This is not slowing down.
This is not theoretical.
This is already happening.

SECTION 4 — The Four AI Superpowers That Actually Matter to Small Businesses

AI can do a thousand things, but you don't need to care about a thousand things.

For small and medium-sized businesses, there are **four capabilities** that change everything.

Superpower #1 — AI Can Understand Information

AI can read and understand:

- emails

- PDFs
- websites
- spreadsheets
- contracts
- transcripts
- chat logs
- text messages

It can pull out key points, spot issues, summarize, prioritize, and organize.

If you're buried in information, this alone is a gamechanger.

Superpower #2 — AI Can Generate Information

AI can create:

- emails and replies
- proposals and quotes
- blog posts and newsletters
- social media posts and ads
- scripts and talking points
- job descriptions
- customer service templates
- meeting summaries
- SOPs and process docs
- training materials

The hardest part of most tasks is starting from a blank page.

AI kills the blank page.

You still review, edit, and approve.
But now you're editing instead of starting from zero.

Superpower #3 — AI Can Analyze Information

AI can analyze:

- financial reports
- sales performance
- marketing metrics
- customer feedback
- survey results
- inventory data
- staff performance metrics (when available)

It can help you answer questions like:

- "Where are we losing money?"
- "Which marketing channels are actually working?"
- "Which customers are most profitable?"
- "Which products or services are underperforming?"

This makes you a smarter, faster decision-maker — without needing a data science degree.

Superpower #4 — AI Can Act on Information

AI can:

- schedule appointments

- send reminders
- route or tag emails
- classify and prioritize leads
- move data between systems
- update CRM records
- assign tasks
- route customers to the right team member
- book calls and follow-ups

This is where **AI + automation** blend into what many people call **AI agents**.

These agents don't just answer questions — they complete tasks.

Soon, AI agents will be the invisible, always-on engine running core parts of modern small businesses.

SECTION 5 — What AI Is NOT (Killing the Most Dangerous Misconceptions)

A lot of the fear around AI comes from half-truths and worst-case scenarios.

Let's clear those out right now.

Misconception #1 — "AI Is Too Complicated"

Truth:
If you can type a sentence, you can use modern AI.

Most AI tools are easier than:

- Excel
- QuickBooks
- Google Ads
- your CRM
- your POS system

You don't need to understand how it works.
You just need to know what to ask it to do.

Misconception #2 — "AI Will Replace All My Employees"

Truth:
AI replaces tasks, not people.

Every credible study shows the same pattern:

People using AI become **more valuable**,
not less.

AI:

- removes repetitive work
- increases throughput
- reduces mistakes
- frees humans for higher-value work

It's a force multiplier, not an automatic pink slip.

Misconception #3 — "AI Will Hurt My Customer Relationships"

Truth:
Customers don't want *more* friction.
They want more:

- speed
- clarity
- reliability

AI helps you:

- respond faster
- avoid dropped messages
- stay consistent
- remember details

Customers care **how well** they are helped, not whether the first response came from a human or an AI-assisted system.

Misconception #4 — "AI Takes Too Long to Learn"

Truth:
You can be meaningfully productive with AI in hours, not months.

If you've ever:

- learned QuickBooks
- figured out a new scheduling app
- set up online banking

- used a CRM

...you've tackled harder tools.

AI is conversation-based.
If you can talk to a person, you can talk to AI.

Misconception #5 — "AI Is Too Expensive"

Truth:
Most SMB AI tools cost less than:

- one hour of professional labor per month
- one lost customer
- one missed opportunity

Typical range: **$20–$100 per month**.

For that, you can:

- save dozens of hours
- eliminate repetitive work
- improve response times
- tighten operations

AI is the **cheapest labor multiplier** in business history.

Misconception #6 — "AI Is Only for Big Companies"

Truth:
If anything, AI benefits small businesses **more**.

Small teams:

- have fewer people
- have more tasks
- have tighter margins
- can't afford to be inefficient

AI lets you:

- do more with less
- punch above your weight
- compete with bigger players

AI doesn't require a massive IT department.
It requires willingness to start small.

Misconception #7 — "My Business Is Too Old-School for AI"

Truth:
AI is already helping:

- law firms
- plumbers
- real estate agents
- dentists and doctors
- home services
- salons and spas
- restaurants and cafes
- accountants and bookkeepers
- landscapers and contractors
- therapists and coaches

- retailers and local shops

This is not a “tech company” issue.
This is a **process and efficiency** issue.

If you use email, appointments, invoices, content, or communication of any kind, AI can help you.

SECTION 6 — The Mental Shift Every Owner Must Make

To succeed with AI, business owners don’t need a tech degree.

They need one mindset shift:

You do not need to understand how AI works.
You only need to understand what it can do for you.

You already live like this:

- You use electricity without knowing the physics.
- You drive a car without understanding the engine.
- You use your phone without understanding satellites.
- You use the internet without knowing how TCP/IP works.

AI is the same.

You do **not** need:

- coding skills
- engineering expertise
- advanced math
- a technical background

You do need to know:

- what problems AI can solve in your business
- how to ask AI the right questions (prompting)
- where AI fits into your existing workflows
- how to measure whether it's working (ROI)
- how to avoid obvious risks

This book will walk you through all of that.

SECTION 7 — What AI Will Become in the Next 3–5 Years (And Why That Matters Now)

If AI already seems powerful, keep this in mind:

The version of AI you're using today is the *worst* it will ever be.

Over the next 3–5 years, you can expect:

- tools to integrate with almost everything you use
- AI agents to complete entire tasks end-to-end
- voice AI that sounds indistinguishable from human staff
- personalization that feels tailored to each customer
- automations that quietly handle entire workflows
- prices that continue to drop
- accuracy and reliability that continue to rise
- AI "co-pilots" inside nearly every business app

The opportunity for early adopters is enormous.
The cost of waiting is brutal.

Business owners who start learning now will:

- adapt faster
- reduce chaos
- improve margins
- build stronger teams
- stay relevant

Those who wait will spend years trying to catch up.

This chapter exists so you don't waste time "wondering what AI is" while your competitors are already using it.

MOVING FORWARD

The Fog Has Lifted — Now You Know What You're Dealing With

At this point, you should have:

- a clear, simple definition of AI
- an understanding of how AI differs from automation and machine learning
- a grasp of why AI keeps getting better (and cheaper)
- A practical view of what AI can do inside a business
- a grounded sense of what AI will **not** do
- fewer misconceptions and more clarity
- a realistic picture of how AI can upgrade your team, not replace it

You don't need to worship AI.
You don't need to fear it.
You just need to understand it well enough to use it.

You're there now.

In the next chapter, we'll tackle the **10 biggest myths** that keep small business owners from taking action — and dismantle them, one by one.

Because clarity is step one.
Courageous action is step two.

You're ready for both.

CHAPTER 3: The 10 Biggest Myths Keeping Small Businesses from Using AI

How Fear, Misinformation, and Outdated Thinking Quietly Kill Growth

INTRODUCTION

Before You Can Use AI, You Have to Unlearn the Lies

Every major technology shift brings confusion.

Electricity.
The automobile.
The internet.
Smartphones.
Social Media.

At first, people argue, dismiss, and misunderstand. Then, slowly, the new thing becomes normal.

AI is different because it feels:

- invisible

- mysterious
- intelligent
- fast-evolving
- unpredictable

And when something feels like that, people don't just misunderstand it — they build **stories** around it.

Most small business owners today are not held back by real technical barriers.
They're held back by:

- "I heard someone say..."
- half-true headlines
- exaggerated horror stories
- assumptions passed around by equally confused peers

In other words: myths.

This chapter blows those myths up.

For each one, we'll spell out:

- **the fear**
- **the truth**
- **a real-world example**
- **what it means for your business**

By the time you're done, there shouldn't be any mental excuses left.

Just clarity, confidence, and a clear path forward.

Let's get into it.

MYTH #1 — "AI Is Too Expensive for Small Businesses."

The Fear:
"AI sounds like something only Google, Amazon, or giant corporations can afford. I run a real business with real bills."

The Truth:
AI is the **cheapest labor multiplier** in business history.

Most AI tools for small businesses cost:

- **$20–$100 per month**
- less than one hour of skilled contractor work
- less than one lost customer due to slow response
- less than the cost of fixing one major mistake

And they regularly save:

- **10–40 hours per month**
- thousands in admin and support costs
- thousands in marketing and content labor
- thousands in operational inefficiencies

Example:
A home services company added a **$29/month AI support tool** to their website and phone system.

Results:

- 18 admin hours saved **per week**
- over **$2,000/month** in labor value reclaimed
- faster response times and happier customers

Total cost? $29.
Total upside? Massive.

If AI feels expensive, it's usually because:

- you haven't calculated the ROI
- you're comparing it to outdated tools
- you underestimate what inefficiency is costing you

Reality:
AI isn't expensive.
Running a modern business without AI is expensive.

MYTH #2 — "AI Will Replace My Whole Team."

The Fear:
"Is this just the first step toward firing people and having robots run my business?"

The Truth:
AI doesn't replace people. It replaces **tasks people shouldn't be doing in the first place.**

AI is ideal for:

- repetitive email writing
- data entry
- scheduling and rescheduling

- basic follow-up
- documentation and note-taking
- inbox sorting and triage
- simple information retrieval

Humans are ideal for:

- building relationships
- selling and negotiating
- solving complex problems
- handling sensitive situations
- creative thinking
- strategy and leadership

AI clears the low-value work so your people can focus on high-value work.

Example:
A dental office used AI to:

- send appointment reminders
- verify insurance details
- handle routine patient questions

The office manager wasn't replaced.
She became **more valuable** because she could now focus on:

- improving billing clarity
- deepening patient relationships
- smoothing scheduling and workflows
- supporting the dentist and team more strategically

Reality:
AI doesn't eliminate jobs.
It eliminates the **worst parts** of those jobs.

MYTH #3 — "I'm Not Tech-Savvy Enough to Use AI."

The Fear:
"I'll break something. I'll click the wrong thing. I'll look stupid. I'm not 'a tech person.'"

The Truth:
Modern AI tools are easier to use than:

- Excel
- email platforms
- QuickBooks
- your POS system
- your CRM
- Meta or Google Ads

If you can **type a sentence**, you can use AI.

Most tools are:

- chat-based
- menu-driven
- guided with prompts
- built for non-technical users

Example:
A 67-year-old florist with **no tech background** uses AI every day to:

- write Instagram captions
- respond to wedding inquiries
- draft proposals
- create bouquet care instructions

Her summary is perfect:

"If you can talk, you can use AI."

Reality:
"Not tech-savvy" is not a real barrier anymore.
It's a story you've outgrown.

MYTH #4 — "AI Will Make My Business Feel Less Personal."

The Fear:
"If I automate responses, customers will feel like I don't care — like they're dealing with a robot instead of a person."

The Truth:
AI, used correctly, makes your business **feel more personal**, not less.

Why?

Because AI helps you:

- respond faster
- remember more
- follow up consistently
- tailor messages to each person
- avoid dropping the ball

Most customers are not craving a long phone call. They are craving:

- accuracy
- clarity
- speed
- respect for their time

AI delivers all four.

Example:
A salon used AI to:

- answer FAQs
- confirm and reschedule appointments
- send reminders and post-visit messages

Customer satisfaction went up **28%.**

Why?

- No more being put on hold.
- No more "we'll call you back" delays.
- Fewer missed appointments.

The stylist still did the hair.
The AI just made everything **around** the service smoother.

Reality:
AI removes friction so your **human moments** can shine.

MYTH #5 — "My Customers Don't Care About AI."

The Fear:
"I run a traditional business. My customers just want human service. They don't care about any of this AI stuff."

The Truth:
Your customers don't care **what** you use. They care **how** they experience you.

Customers want:

- fast replies
- clear answers
- simple booking
- reliable follow-up

AI gives you all of that.

Nobody is thinking,
"I hope a human is the one who sends my appointment reminder."

They're thinking,
"I hope someone responds quickly so I can move on with my day."

Example:
A home inspector added AI chat to their website to:

- answer questions
- schedule inspections
- provide prep instructions

Within 60 days:

- bookings **doubled**
- customers repeatedly commented:
  "I loved how quickly you got back to me."

Reality:
Your customers already expect AI-level speed and convenience — whether you use AI or not.

MYTH #6 — "AI Will Give Wrong Answers and Embarrass My Business."

The Fear:
"What if AI says something wrong, weird, or off brand? I can't risk that with my clients."

The Truth:
Yes, AI can be wrong. So can humans.

The businesses that get burned by AI usually:

- let AI respond without any oversight in high-risk areas
- give AI access to the wrong things
- use it without guardrails or guidelines
- treat it like a replacement instead of an assistant

When used properly, AI is actually **more accurate and consistent** than humans at:

- scheduling
- data entry
- drafting responses for review

- summarizing information
- following rules and templates
- documentation

Example:
A real estate brokerage used AI to **draft emails**, but agents always reviewed before sending.

Results:

- email error rate dropped from **13% to 3%**
- tone and messaging improved
- follow-up became more consistent

The AI didn't embarrass them.
It made them look **more professional.**

Reality:
AI plus human review is safer than human-only work in most admin and communication tasks.

MYTH #7 — "AI Takes Too Long to Set Up."

The Fear:
"I don't have time for another massive tech project. I'm already behind on everything."

The Truth:
Most high-impact AI tools for Small and Midsize Businesses (SMBs) are:

- installed in **10–20 minutes**
- learned in **1–2 hours**
- delivering value, the **same day or same week**

You're not rebuilding your infrastructure.
You're bolting on upgrades.

Example:
A coffee shop installed an AI-powered ordering and FAQ assistant in one afternoon.

By the next morning:

- staff phone time dropped **40%**
- fewer interruptions at the register
- more accurate orders
- happier team and customers

AI is fast to set up because:

- It's designed for non-technical users
- most SMB tools are no-code
- integrations often happen with a few clicks

Reality:
"It takes too long to set up" is almost always an excuse — not a fact.

MYTH #8 — "AI Will Cause Legal or Compliance Problems."

The Fear:
"What if we accidentally violate a law, break confidentiality, or mishandle data by using AI?"

The Truth:
Used recklessly, any tool can create risk.
Used wisely, AI can actually **reduce** your risk.

With basic safety practices (which we'll cover in detail in Chapter 14), AI can help:

- standardize documentation
- reduce data-entry errors
- keep better records
- improve consistency of messaging
- prevent things from falling through the cracks

Example:
A physical therapy clinic used AI to help generate treatment summaries and documentation drafts for staff to review.

Impact:

- documentation accuracy improved
- staff spent less time writing notes
- compliance risk **decreased**

AI didn't create new legal problems.
It helped them do the right things more consistently.

Reality:
AI, used with clear policies and human oversight, is **safer** than manual, error-prone processes.

MYTH #9 — "AI Is a Fad — It'll Blow Over Like 3D TVs."

The Fear:
“This feels like a trend. Maybe I can wait it out and keep doing things the way I always have.”

The Truth:
AI is not a gadget.
It’s **infrastructure.**

It belongs in the same category as:

- electricity
- the internet
- smartphones
- cloud software

It’s becoming:

- the new communication layer
- the new productivity layer
- the new administrative layer
- the new customer experience layer
- the new hiring and operations layer

And it’s already baked into:

- Microsoft 365
- Google Workspace
- Shopify
- QuickBooks and Xero
- HubSpot and Salesforce

- Stripe
- Adobe
- Canva
- countless CRMs, POS systems, and support tools

This isn't something you opt into.
It's something that is becoming **standard.**

Reality:
You will use AI. The question is:
Will you use it early, for advantage — or late, just to survive?

MYTH #10 — "If I Use AI Wrong, It Could Destroy My Business."

The Fear:
"If I mess this up, I could lose customers, hurt my reputation, or break something important."

The Truth:
Can AI be used badly? Of course. So can email, phones, websites, and employees.

But small businesses that win with AI don't start with risky, high-exposure experiments. They:

- start small
- test in low-risk areas
- keep humans in the loop
- use reputable tools
- follow straightforward guidelines

- expand only after early wins

The biggest threat is not “using AI wrong.”

The biggest threat is falling so far behind that you never catch up.

Example:
Two similar insurance agencies:

- **Agency A** used AI to handle basic follow-up, quote reminders, and document drafts.
- **Agency B** kept doing everything manually and “waited to see how things played out.”

Twelve months later:

- Agency A grew revenue by **27%**
- Agency B declined **18%**

Within a few years, Agency B admitted:

“We didn’t lose to AI.
We lost to competitors who used it while we hesitated.”

Reality:
Using AI carefully and conservatively is far **less dangerous** than ignoring it.

The Real Reason These Myths Exist

Small business owners don’t avoid AI because they’re bad with technology.

They avoid AI because they’re afraid of:

- looking incompetent
- making a visible mistake
- being embarrassed in front of customers or staff
- losing control over their processes
- admitting they don't fully understand something
- dealing with yet another change
- confirming the feeling that they're "behind"

Those are not tech fears.
They are **human** fears.

This book is designed to walk you through those fears — not by shaming you, but by giving you:

- clarity
- simple steps
- realistic expectations
- real-world examples
- safe starting points

You don't need to pretend you're not worried.
You just need to be more committed to your business's future than to your comfort.

What Happens When SMBs Finally Break These Myths

When owners stop believing these myths and start experimenting, three things happen quickly:

1. They Get HOURS of Their Time Back

AI quietly wipes out:

- routine admin
- repetitive customer responses
- manual documentation
- basic marketing production
- painful follow-up tasks

Suddenly:

- inboxes feel lighter
- to-do lists shrink
- evenings and weekends start to reappear

2. They Regain Confidence as Leaders

What used to feel overwhelming becomes manageable.

Instead of:

- "I don't get this stuff,"

it becomes:

- "I know where AI fits, what we're testing, and what it's doing for us."

They move from **defensive** to **proactive.**

3. They See Growth Faster Than Expected

Because AI accelerates:

- revenue (better follow-up, better offers, better content)
- retention (better service, fewer dropped balls)
- productivity (more output, fewer mistakes)
- capacity (more work with the same team)

The business feels lighter, stronger, and more scalable.

This chapter exists to clear the mental runway.

The rest of the book exists to help you **take off.**

MOVING FORWARD

The Myths Are Gone — Now the Real Transformation Can Begin

In the first three chapters, you have:

- seen the divide between AI Adopters and Holdouts
- learned what AI really is and how it works in plain English
- dismantled the biggest myths keeping business owners frozen

You now have:

- clarity on the playing field
- a grounded understanding of AI
- fewer excuses and more options

Part I ends here — with a new mindset.

Next, we move from **beliefs** to **action**:

- real opportunities
- specific workflows
- recommended tools
- financial impact
- a step-by-step path to becoming an AI-enabled business

You're done warming up.

Now we build.

CHAPTER 4: The Overwhelm Problem: "I Don't Know Where to Start"

Why Most Small Businesses Freeze Before They Ever Use AI — and How to Break Through It in One Simple Step

INTRODUCTION

The Real Enemy Isn't AI. It's Overwhelm.

If you're like most small-business owners, it's not that you don't want to use AI.
It's not that you don't understand its value.
It's not that you're anti-technology.

It's this:

You simply don't know where to start.

The modern AI landscape is loud, chaotic, and crowded.
Everyone is pushing a new tool.
Every week there's a new "must-have" system.
Every influencer swears their method is the answer.
Every platform promises automation, growth, and time savings.

Small-business owners don't lack intelligence or willingness. They lack a **clear, simple, practical starting point**.

The truth is this:

Overwhelm—not cost, not fear, not lack of knowledge—is the #1 reason small businesses fail to adopt AI.

Overwhelm freezes owners.
Frozen owners delay action.
Delayed action becomes lost opportunity.
And lost opportunity becomes falling behind.

This chapter fixes that problem completely.

You're about to learn the simplest, cleanest, most effective way to start using AI in your business—without stress, without complexity, and without disrupting your entire operation.

This chapter is your bridge from confusion to clarity, from hesitation to momentum, and from overwhelm to meaningful results.

Let's begin.

SECTION 1 — Why Overwhelm Is More Dangerous Than Ignorance or Fear

A business owner who doesn't understand AI can be taught.

A business owner who fears AI can be reassured.

But a business owner who is overwhelmed?

They freeze.
They hesitate.
They spin in circles.
They talk about AI for months.
They read articles but take no action.
They attend workshops but implement nothing.
They save links, take notes, and never start.

Overwhelm is paralysis disguised as "thinking" or "planning."
It kills more innovation than cost.
It kills more growth than competition.
It kills more progress than failure ever will.

And here's the harsh truth:

Overwhelm is not the result of a lack of ability.
It's the result of too many options and no roadmap.

Which is exactly what we solve next.

SECTION 2 — The Three Causes of AI Overwhelm

Most business owners aren't overwhelmed because AI is complicated.
They're overwhelmed because the ecosystem around AI *is*.

Every overwhelmed owner falls into at least one of these three categories:

1. Choice Overload

There are more than **10,000+ AI tools** in the market.

Every day, there are new launch posts, new demos, new lists:

- “Top 50 AI tools you need.”
- “The AI tool you MUST use in the next 90 days!”
- “The secret AI system no one is talking about!”

It’s impossible to evaluate everything.
It’s impossible to know what’s legitimate.
It’s impossible to know what matters.

So, owners quietly think:

“I’ll wait until this becomes more clear.”

But clarity never comes on its own.

2. Priority Confusion

Most owners try to fix everything at once:

- “I need AI for marketing.”
- “I also need AI for customer service.”
- “And for hiring.”
- “And for content.”
- “And maybe for bookkeeping...”

AI becomes a giant project instead of a simple tool.

Here’s the hard truth:

Trying to automate your entire business at once guarantees failure.

3. Fear of the Wrong Decision

Behind most hesitation is a quiet internal fear:

- “What if I choose the wrong tool?”
- “What if this breaks something?”
- “What if this creates a mess?”
- “What if I waste money?”
- “What if I don’t understand it?”

This fear creates what psychologists call *avoidance patterns*:

- Checking reviews instead of deciding
- Watching videos instead of experimenting
- Bookmarking resources instead of acting
- Writing lists instead of choosing one solution
- Saying “soon” instead of now

And AI becomes the diet people start “next Monday”—forever.

SECTION 3 — The Real Psychology Behind Business Owner Overwhelm

Most owners won’t admit this, but they think:

“I should already understand this.”

So when they don’t, they feel embarrassed, frustrated, or insecure.

That embarrassment is worse than the confusion itself.

They avoid starting because starting feels like admitting they’re behind.

And here is where your thinking must shift:

You are not behind.
You are early—if you start now.

No business owner is supposed to understand AI intuitively.
You didn't understand payroll intuitively.
You didn't understand taxes intuitively.
You didn't understand advertising intuitively.

You learned those things because you had to.

AI is the same.

This chapter gives you the simplest learning path possible.

SECTION 4 — The Solution: The Single Workflow Method

Every successful small-business AI implementation begins with this rule:

Start with ONE workflow. Not tools. Not systems. ONE workflow.

Instead of asking:

"Which tools should I use?"
"Where do I begin?"
"What should I automate first?"
"Which department needs AI most?"

Ask this:

“What is one repetitive, high-friction, time-wasting workflow in my business that costs me money or frustrates customers?”

Examples:

- Missed or slow follow-up
- Customer questions
- Appointment scheduling
- Social content
- Repetitive emails
- Invoicing
- Staff reminders
- Estimate creation
- Inventory descriptions
- Document drafting

If a task:

- happens weekly
- takes effort
- affects customers
- affects money
- requires writing
- is repetitive

...it is a perfect AI starting point.

This ONE workflow determines which tool you choose.
This ONE workflow determines your first AI win.
This ONE workflow removes overwhelm completely.

You go from:

"I don't know where to start"
→ to
"I know exactly where to start."

SECTION 5 — How to Choose Your First AI Workflow

Use the **Two-of-Three Rule**:

Your starter workflow must satisfy at least 2 of these 3:

1. High Friction

This task annoys you or your team.
It's slow, messy, or chaotic.

2. High Repetition

It happens every day or every week.

3. High Cost

It eats time, loses money, or frustrates customers.

If your workflow checks two of these boxes → **start there**.

SECTION 6 — Choosing Your First AI Tool (The Four Rules)

Once you know the workflow, the tool becomes obvious.

Apply these four rules:

Rule #1: The tool must solve ONE workflow.

Not ten. Not five. One.

Rule #2: The tool must be easy.

If you can't set it up in under an hour, it's too complicated.

Rule #3: The tool must cost less than $50/month.

Unless it directly increases revenue.

Rule #4: You should feel momentum within 7 days.

If not? Switch tools.

AI success is not about choosing the perfect tool.
It's about choosing a simple tool that solves a real problem quickly.

SECTION 7 — The 30-Day AI Starter Plan (Simple, Stress-Free, Proven)

This is the same plan I use with consulting clients.
It works for gyms, med spas, HVAC companies, boutiques, real estate teams, dentists, and contractors.

WEEK 1 — Diagnose Your Bottleneck

Ask these three questions:

- Where am I losing the most **time**?
- Where am I losing the most **money**?
- Where am I frustrating **customers**?

Pick the workflow that fits at least two answers.

This is your starting point.

WEEK 2 — Choose ONE AI Tool

Examples:

For communication & writing

ChatGPT, Claude, Jasper

For customer responses

Tidio, GoHighLevel, Intercom, Air.ai

For admin

Zapier, Notion, Google Workspace AI

For scheduling

Calendly automation

For finance

QuickBooks AI, Wave AI

Choose **one** tool only.

WEEK 3 — Implement a Real System

- Connect the software.
- Train one team member.
- Build one automation.

- Test with real customers.
- Refine based on feedback.

Start with one simple outcome:

- Auto-reply
- Auto-schedule
- Auto-write
- Auto-summarize.
- Auto-invoice
- Auto-update

Momentum > perfection.

WEEK 4 — Measure Success

Ask:

- Did this workflow become easier?
- Did it save time?
- Did it reduce errors?
- Did customers notice?
- Did it save or earn money?

If yes → move on to workflow #2.
If no → adjust and re-test.

The system grows from here—naturally, without overwhelm.

SECTION 8 — Case Study: The Boutique That Was Stuck but Became Unstoppable

A two-person boutique faced:

- inconsistent social media
- slow customer replies
- weak product descriptions
- inconsistent marketing
- owners working late nights.

They were overwhelmed and embarrassed.

We applied the **Single Workflow Method**.

Chosen Workflow: **Content Creation**
Tools: **ChatGPT + Canva AI**

In 30 days:

- 65 hours saved
- 30 days of scheduled posts
- 22% increase in foot traffic
- $2,400 month-over-month growth
- Owners regained evenings.
- Consistency went from zero to professional.

Their business didn't transform because they used dozens of tools.
It transformed because they automated **ONE** workflow.

SECTION 9 — What Success Looks Like After 30 Days

Small businesses that adopt AI using this method consistently report:

- More time
- Less stress
- Better communication
- Faster workflows
- More consistent marketing
- More confident teams
- Immediate measurable ROI

AI becomes part of the business naturally—because it started simply.

SECTION 10 — The Mistakes to Avoid

X Trying to automate everything
X Choosing complex tools
X Starting with sensitive data
X Training the whole team at once
X Not testing with real customers
X Expecting perfection
X Switching tools constantly
X Buying too many subscriptions

Avoid these, and your first AI win will come fast.

SECTION 11 — The Mindset Shift That Changes Everything

You are not trying to "implement AI."

You are trying to **fix one workflow**.

AI is not a project.

AI is not a department.

AI is not a massive undertaking.

AI is a tool—like a calculator, a phone, or email.
You don't "implement email."
You send your first message.

AI works the same way.

MOVING FORWARD

You're One Workflow Away from Momentum

The biggest lie in the AI industry is that adoption must be massive, expensive, or overwhelming.

The truth is far simpler:

Start small. Win fast. Scale gradually.

If you follow the Single Workflow Method, you will build momentum.
Momentum builds confidence.
Confidence builds capability.
Capability builds transformation.

This chapter was designed to remove the biggest barrier in your way—overwhelm.

Now you know exactly where to begin.

Turn the page.

Your AI-powered business starts now.

CHAPTER 5: Your First 30 Days With AI: A Simple, Stress-Free Blueprint for Small Businesses

How to Build Momentum, Reduce Overwhelm, and Create Your First Real AI Win

INTRODUCTION

The Most Dangerous Mistake Small Businesses Make with AI

Most small business owners make the same fatal mistake when they begin exploring AI:

They try to overhaul **everything** at once.

They try to:

- automate customer service
- rewrite their marketing
- fix their hiring problems
- build new content systems
- streamline operations
- implement new software

- reorganize their entire workflow

All in the same week.

And what happens?

They burn out.
They quit.
And they walk away convinced that “AI doesn’t work.”

Not because AI failed them —
but because they tried to eat the whole elephant in a single bite.

This chapter protects you from that spiral.

You’re about to learn the **Single Workflow AI Method**, your proven, beginner-safe, 30-day plan for launching AI inside your business **without stress, confusion, or chaos**.

By the time you finish this chapter, you’ll know:

- exactly which workflow to automate first
- which tool to choose (without shopping for 20)
- how to implement AI safely
- how to measure what’s working
- how to produce real ROI within 30 days

This is the exact system I would use to teach:

- HVAC companies
- med spas
- salons

- real estate teams
- accountants
- restaurants
- construction companies
- e-commerce stores
- local service businesses

It works because it's simple.
It works because it's grounded in your everyday reality.
It works because it's built to eliminate overwhelm and create momentum.

Let's begin.

SECTION 1 — Why the First 30 Days Matter More Than the First 12 Months

AI adoption is not a talent issue.
It's not a budget issue.
It's not a tech background issue.

It is a **momentum issue**.

Your first 30 days determine:

- how confident you feel using AI
- how quickly your team buys in
- whether you see fast ROI
- whether you continue adopting AI
- or stop completely

Businesses that follow this plan almost always expand their AI use.

Businesses that skip it?
They drown in complexity and never get off the ground.

The truth is simple:

You don't need to start big.
You need to start correctly.

SECTION 2 — The Single Workflow AI Method

Your Antidote to Overwhelm

Forget the idea of adopting 10 tools.

Forget the idea of automating your whole business.

You're going to automate **one workflow**.

This method is the gold standard because it:

- reduces overwhelm
- builds early wins
- speeds up team adoption
- protects you from costly mistakes
- maximizes ROI
- teaches you how AI behaves
- proves AI value before wider rollout

It is simple.
It is powerful.

And it is the secret behind every successful AI transformation I've seen.

SECTION 3 — Your First 30 Days With AI (Step-by-Step)

Here is your complete, stress-free, editor-approved blueprint.

WEEK 1 — Diagnose Your Business Bottleneck

Your mission in Week 1:

Choose ONE workflow that represents your biggest pain point.

We are applying the **One-Workflow Rule**:
Your starting point must be:

- high friction
- high repetition
- high cost (time or money)

Start by asking three diagnostic questions:

1. Where am I losing the MOST time?

Examples:

- replying to customers
- writing emails
- scheduling appointments
- creating social content
- billing and invoicing
- writing quotes or estimates

- paperwork
- documentation

2. Where am I losing the MOST money?

Examples:

- missed leads
- slow follow-up
- delayed invoicing
- pricing errors
- admin inefficiencies
- manual data entry
- no-shows or cancellations

3. Where am I frustrating customers?

Examples:

- slow responses
- inconsistent communication
- errors in scheduling
- long wait times
- confusing processes
- outdated website info
- dropped messages

Your Best Starting Workflow Will Hit at Least Two of These Three.

The "Pick One" Rule

Choose the workflow that:

- takes too much time
- happens weekly or daily
- is mostly language-based (AI's sweet spot)
- affects customers or revenue
- does NOT involve sensitive data (yet)
- is easy to explain

Do **NOT** choose:

- complex, multi-system processes
- low-frequency tasks
- workflows involving legal/medical/financial risk
- workflows nobody on your team understands
- workflows you haven't clearly mapped

Start simple.
Start measurable.
Start safe.

WEEK 2 — Choose ONE Tool (Not Six)

The fastest way to sabotage your first month with AI is to sign up for every shiny tool you find.

Don't do that.

You pick ONE tool based on the workflow you chose.

Here's the exact short list you should choose from — nothing else.

If your workflow involves writing or communication:

- ChatGPT
- Claude
- Jasper
- Canva AI

If your workflow involves customer response or appointments:

- Tidio
- Intercom
- GoHighLevel
- Air.ai
- Smith.ai

If your workflow is admin or repetitive tasks:

- Zapier
- Make (Integromat)
- Notion AI
- Google Workspace AI

If your workflow is bookkeeping or finance:

- QuickBooks AI
- Wave AI
- Xero AI

Price Rule:

Your first tool should cost **under $50/month**
unless the tool directly increases revenue (e.g., GoHighLevel).

Do not compare 20 tools.
Do not watch 11 YouTube tutorials.
Do not get lost in reviews.

Compare **three**, pick **one**, move on.

WEEK 3 — Build a Small, Real AI System

(Where Most Businesses Fail)

This is the week where most owners struggle.

They sign up for a tool...
and never integrate it.

They poke around the dashboard...
and never build a workflow.

They "mean to get back to it..."
and forget.

Not you.

Here's your job in Week 3:

1. Set up the tool COMPLETELY.

Not halfway.
Not "I'll finish later."

You will:

- connect accounts
- integrate your website, CRM, or email
- set up basic automations
- upload templates
- define rules
- test the workflow end-to-end

2. Train ONE employee — not the whole staff.

Pick:

- your fastest learner
- your most curious team member
- or the person who does the workflow today

Do NOT roll out AI company-wide yet.

3. Automate ONE outcome.

Examples:

- auto-reply to new leads
- auto-schedule appointments
- auto-summarize emails
- auto-generate weekly social posts
- auto-tag customer inquiries
- auto-generate invoices or receipts
- auto-draft proposals

ONE output.

Not five.
Not a whole department.

One.

Mastering one output builds confidence and wins support from your team.

WEEK 4 — Measure ROI and Decide What Happens Next

This is where the magic happens.

In Week 4, you evaluate the workflow you built.

Measure:

- hours saved
- labor cost saved
- new leads captured
- response time improvements
- errors reduced
- revenue influenced
- customer satisfaction
- employee feedback

Then ask these three questions:

1. Did this workflow get easier?

If YES → Continue and expand.
If NO → Adjust and try again.

2. Did this tool save or make money?

If YES → Keep it.
If NO → Replace it.

3. Did customers or employees notice improvements?

If YES → You've found a winner.
If NO → Refine the automation or prompt.

Your Goal Is NOT Perfection.

Your Goal Is MOMENTUM.

Your goal is to build your first AI win — and build it fast.

This early win becomes the foundation for every AI upgrade you introduce next.

Case Study: The Med Spa That Got 2 Extra Staff Hours Per Day

A med spa with a four-person team was overwhelmed by:

- nonstop phone calls
- appointment changes
- repetitive questions
- new customer inquiries
- lead follow-up delays

They couldn't hire another receptionist — labor costs were too high.

We applied the Single Workflow AI Method:

Chosen Workflow:

Appointment handling
(inquiries → scheduling → reminders)

Chosen Tool:

Tidio

Implementation:

Custom AI receptionist built in one afternoon.

Results in 30 Days:

- 57% fewer phone calls
- 99% of leads responded to within 60 seconds
- owner saved 10 hours per week
- staff reassigned to higher-value interactions
- 11% revenue increase in 60 days
- zero customer complaints
- cost = $29/month

This is why starting with **ONE** workflow works.

Small steps → fast wins → big confidence → real ROI → long-term transformation.

MOVING FORWARD

Your First 30 Days Determine Your AI Future

AI success is not built on complexity.
It is built on clarity.

And nothing creates clarity faster than:

- **one workflow**
- **one tool**
- **one month**
- **one early win**

This is your foundation.

Once you've completed your first 30 days, you'll have:

- real results
- confidence
- team buy-in
- measurable ROI
- momentum
- a clear next step

Now you're ready for what comes next:

- ➢ **Choosing your second workflow**
- ➢ **Integrating AI into multiple departments**
- ➢ **Scaling your systems**
- ➢ **Building your AI-Enabled Business**

Your transformation begins with one small, smart, strategic move.

You've got your blueprint.

Now you're ready for your breakthrough

CHAPTER 6: The SMB AI Tech Stack

10 Essential Tool Categories Every Small Business Should Consider (And How to Use Them Without Wasting Money)

INTRODUCTION

Your Tech Stack Will Decide Whether You Compete or Get Crushed

Most small business owners don't think of themselves as "tech companies."

You're a:

- restaurant
- home services company
- medical office
- boutique
- gym
- law practice
- construction firm

You didn't start your business because you love software.
You started it because you love the work.

But here's the hard truth:

In the age of AI, every business is a technology business — whether you admit it or not.

That doesn't mean you need to:

- learn to code
- understand machine learning
- know how neural networks are trained

It does mean this:

The **software you choose** — your **AI tech stack** — will determine:

- how fast you grow
- how profitable you become
- how hard or easy your life is as an owner
- how efficient your employees can be
- how customers experience your brand
- how well you handle hiring, scheduling, and operations
- how resilient your business is when things get hard

Your tech stack is no longer "back office."
It is your competitive foundation.

The gap between businesses with a smart, lean AI tech stack and those without it is widening **every month.**

This chapter will walk you through:

- the **10 core categories** of AI tools every SMB should be aware of
- what each category actually does in the real world
- how to choose tools **without overspending or overcomplicating** your life
- the kind of ROI each category should deliver
- specific workflow examples by industry
- the biggest mistakes to avoid
- a simple **SMB AI Stack Blueprint** you can copy

By the end, you'll know **exactly** which tools belong in your business — and which you can happily ignore.

Let's build the backbone of your AI-powered company.

SECTION 1 — What an AI Tech Stack Is (And What It Isn't)

When business owners hear "tech stack," they usually think:

- expensive software
- complicated integrations
- consultants
- disruption
- risk
- time they don't have

But a **tech stack** is not some giant enterprise IT diagram.

A tech stack is simply:

The systems your business uses to communicate, operate, market, sell, schedule, fulfill, and deliver.

Your **AI tech stack** sits on top of, and inside, those systems and:

- speeds them up
- automates repetitive work
- reduces human error
- supports your team
- responds to customers faster
- gives leadership better visibility

A modern AI tech stack has **four jobs**:

1. **Automate low-value tasks**
   Stop paying humans to do work machines can do better, faster, cheaper.

2. **Augment your team**
   Give every employee the equivalent of a tireless assistant.

3. **Create consistency**
   In marketing, customer responses, operations, hiring, documentation.

4. **Improve decision-making**
   With clearer reporting, forecasting, and real-time insights.

The tools in this chapter are not toys.
They are the building blocks of those four outcomes.

SECTION 2 — The 10 Essential Categories of an SMB AI Tech Stack

You do **not** need every tool in every category.

But you should understand each category well enough to say:

- "We need this now."
- "We might need this later."
- "We can skip this completely."

Here are the 10 core categories nearly every small business should at least consider:

#	**Category**	**Examples**	**Primary Purpose**
1	AI Writing & Knowledge Tools	ChatGPT, Claude, Gemini, Perplexity	Content, communication, training, research
2	AI Customer Response & Appointment	Tidio, Intercom, Air.ai, Smith.ai	Instant replies, booking, 24/7 customer service
3	AI CRM & Sales Automation	HubSpot, GoHighLevel, Close	Lead tracking, follow-up, pipelines

#	Category	Examples	Primary Purpose
4	AI Marketing & Content Creation	Canva AI, Jasper, Adobe Firefly, Descript	Social, ads, video, design
5	AI Operations & Workflow Automation	Zapier, Make	Connecting tools, automating processes
6	AI Scheduling & Appointment Tools	Calendly, TidyCal	Booking, reminders, calendar management
7	AI Bookkeeping & Finance	QuickBooks AI, Wave AI, Xero AI	Accounting, billing, cash flow
8	AI HR & Hiring	BambooHR AI, Rippling AI, Homerun AI	Recruiting, screening, onboarding
9	AI Knowledge & SOP Management	Notion AI, Guru AI	SOPs, training, internal knowledge base
10	Industry-Specific AI Tools	Shopify Magic, Toast AI, Autodesk AI	Deep, specialized workflows

Now we'll break down each category — what it does, who it's for, and how not to waste money on it.

CATEGORY 1 — AI Writing & Knowledge Tools

ChatGPT, Claude, Gemini, Perplexity

Why This Category Matters

This is the **foundational** AI category for almost every business.

These tools act as:

- your writing assistant
- your researcher
- your marketing intern
- your customer communication coach
- your analyst
- your SOP generator
- your training manual writer

They quietly upgrade almost every written communication in your company.

Critical Use Cases

- Writing and replying to emails
- Drafting marketing content (newsletters, blogs, ads)
- Creating proposals and quotes
- Writing job descriptions and hiring ads
- Drafting SOPs, manuals, and policies

- Summarizing long documents (contracts, reports, transcripts)
- Generating FAQ content for your website
- Creating phone scripts or front-desk templates
- Doing market and competitor research
- Brainstorming campaign ideas

Industry Examples

- **Home Services:** Estimate templates, follow-up emails, script for technicians.
- **Med Spas/Clinics:** Patient communication drafts, aftercare instructions, website copy.
- **Restaurants:** Menu descriptions, promotions, event announcements, social posts.
- **Boutiques:** Product descriptions, email campaigns, social captions.
- **Professional Services:** Proposals, client summaries, reports, presentations.

How to Choose

- **ChatGPT** → Most versatile, strong all-rounder.
- **Claude** → Excellent writing quality, great for long-form and nuanced tone.
- **Gemini** → Best if you live in Google Workspace (Docs, Sheets, Gmail).

- **Perplexity** → Best for research-heavy questions and citing sources.

Big Mistake to Avoid

Using these tools **randomly**.

You need a **prompt library**:

- standard prompts for emails
- standard prompts for blog posts
- standard prompts for proposals
- standard prompts for SOPs

A small set of well-tested prompts will give you **10x more value** than poking around with new prompts every day.

We've provided you with twenty five key prompts you can use immediately to begin to use AI in your business more effectively.

CATEGORY 2 — AI Customer Response & Appointment Automation

Tidio, Intercom, Air.ai, Smith.ai

Why This Category Matters

This is one of the **highest-ROI categories** for small businesses.

Most SMBs lose:

- 40% of leads to slow or missed responses
- customers who bounce if they don't hear back quickly

- thousands per year from poor follow-up and bottlenecks

AI-powered customer response tools:

- answer instantly
- capture leads 24/7
- book appointments
- reduce phone volume
- qualify leads

Critical Use Cases

- Instant chat replies on your website
- Automated answers to common questions
- Booking or rescheduling appointments
- Capturing name, email, and phone number before a human gets involved
- Handling hours, location, pricing FAQs
- Reducing basic phone calls and emails
- Routing urgent issues to humans

Industry Examples

- **Med Spa:** 24/7 AI receptionist that answers FAQs and books appointments.

- **HVAC Company:** AI answers "Do you serve my area?" and "Can you come today?" instantly.
- **Real Estate Team:** AI pre-qualifies buyers before an agent calls.
- **Restaurant:** AI handles reservation requests, hours, menu questions.

Choosing the Right Tool

- **Tidio** → Affordable, easy, great for chat + basic automation.
- **Intercom** → Strong for scaling companies with lots of inbound messages.
- **Air.ai** → Voice-based AI agents for phone calls.
- **Smith.ai** → AI + human receptionists for a more "white glove" experience.

Pitfall to Avoid

Using a bot that sounds robotic and wrong.

Fix this by:

- feeding it a short "Company Facts" document
- giving it tone guidelines (friendly, professional, casual, etc.)
- testing it yourself every week

CATEGORY 3 — AI CRM & Sales Automation

HubSpot, GoHighLevel, Close, Salesforce (SMB)

Why This Category Matters

Every business lives or dies by:

- generating leads
- following up consistently
- closing and retaining customers

An AI-enhanced **CRM** (Customer Relationship Management) system is the **engine** behind this.

Good CRMs with AI can:

- score leads automatically
- recommend who to contact next
- generate follow-up emails and texts
- log calls and messages
- track the customer journey
- trigger sequences based on behavior

Critical Use Cases

- Automated email/SMS follow-up
- Lead scoring (who is likely to buy)
- Pipeline tracking and reminders
- Proposal and quote generation
- Win-back campaigns for old leads
- Upsell and renewal reminders
- Abandoned cart follow-ups (ecommerce)

How to Choose

- **HubSpot:** Great free tier, intuitive, solid for many SMBs.
- **GoHighLevel:** All-in-one powerhouse for agencies and service businesses.
- **Close:** Awesome for phone-heavy sales teams.
- **Salesforce SMB:** Very powerful but often overkill for small teams.

Big Mistake to Avoid

Buying a CRM and then **never building automations**.

If your CRM isn't sending automatic follow-ups, reminders, and nurture messages, you're using a race car like a bicycle.

CATEGORY 4 — AI Marketing & Content Creation

Canva AI, Jasper, Adobe Firefly, Descript

Why This Category Matters

Marketing is the **chronic pain point** of small businesses.

Most owners:

- post inconsistently
- hate staring at blank screens
- don't have designers
- don't have copywriters
- have no time

AI marketing tools:

- create graphics
- draft ad copy
- propose content calendars
- write emails
- script videos
- generate images and visuals

Critical Use Cases

- Social media posts and carousels
- Graphics, flyers, one-sheets
- Email campaigns and newsletters
- Blog posts & SEO content
- Product descriptions
- Short-form video scripts (Reels, TikTok, Shorts)
- Landing page content
- Event promotions

Best Tools

- **Canva AI:** Design, social media, branding, templates.
- **Jasper:** Marketing-focused copywriting.
- **Adobe Firefly:** Brand-safe image generation.
- **Descript:** AI video editing, captions, and repurposing.

Industry Examples

- **Real Estate:** Listing descriptions, social graphics, open house ads.
- **Restaurants:** Daily specials, event graphics, promo campaigns.
- **Boutiques:** Product drops, lookbooks, seasonal campaigns.
- **Gyms:** Challenge promotions, transformation posts, testimonials.

Major Warning

If you let AI write generic "blah" content, people will tune you out.

You must:

- feed it your brand story
- show it examples of your tone
- build branded prompts ("Write this like [your brand]: friendly, local, no cliché hype.")

CATEGORY 5 — AI Operations & Workflow Automation Engines

Zapier, Make

Why This Category Matters

If AI is the **brain**, automation tools are the **nervous system**.

These tools connect all your systems so:

- when something happens in one place,
- the right things happen everywhere else
- automatically

Common Use Cases

- When someone fills out a form → add to CRM → send welcome email → notify team.
- When someone books an appointment → add to Google Calendar → send confirmation → send reminder.
- When an invoice is paid → update QuickBooks → send receipt → mark as paid in CRM.
- When a lead clicks a link → start a follow-up sequence.
- When a contact cancels → remove from sequences → tag for win-back later.

Best Tools

- **Zapier:** Easiest for most small businesses, tons of integrations.
- **Make:** More powerful and flexible, better at scale.

Warning

Don't start by automating high-risk workflows (compliance, medical, legal).
Start with **simple, reversible automations** that save time and reduce manual entry.

CATEGORY 6 — AI Scheduling & Appointment Automation

Calendly, TidyCal

Why This Category Matters

Manual scheduling is quiet chaos:

- back-and-forth emails
- missed calls
- no-shows
- double bookings

AI-boosted scheduling tools:

- let customers book themselves
- send reminders
- reduce no-shows
- handle time zones
- route bookings to the right person

Perfect For:

- coaches and consultants
- med spas and clinics
- salons and barbershops
- real estate agents
- trainers and gyms
- law firms
- accountants and tax pros
- home service estimates

Recommended Setup

- **Calendly** with workflows and reminders
- Integrated with your CRM & calendar via Zapier/Make

CATEGORY 7 — AI Bookkeeping & Finance

QuickBooks AI, Wave AI, Xero AI

Why This Category Matters

Most small businesses are weaker than they should be financially because they:

- delay bookkeeping
- miscategorize transactions
- send invoices late
- don't look at cash flow until there's a crisis

AI-enhanced finance tools can:

- auto-categorize expenses
- spot anomalies
- generate P&Ls and basic forecasts
- remind customers about overdue invoices
- produce summaries for you and your accountant

Use Cases

- "Summarize this month's finances in plain English."
- "Show me which customers are past due."
- "Predict my cash position 30 days from now."
- "Generate a simple financial report for my bank."

They won't replace your accountant.
They'll make your accountant — and your decisions — better.

CATEGORY 8 — AI HR, Hiring & Staff Management

BambooHR AI, Rippling AI, Homerun AI

Why This Category Matters

Hiring is painful for most SMBs.

You:

- get buried in resumes
- write generic job posts
- waste time interviewing the wrong people
- have weak onboarding

AI can:

- write sharper job posts
- screen resumes for basic fit
- rank candidates
- draft interview questions
- generate onboarding docs
- help with basic training content

This category becomes more valuable as you grow beyond a handful of employees.

CATEGORY 9 — AI Knowledge & SOP Management

Notion AI, Guru AI

Why This Category Matters

Every business runs on **tribal knowledge**:

- "Ask Sarah, she knows how."
- "That's in some old Google Doc."
- "I think we changed that last year?"

AI knowledge tools turn your business into a **self-documenting, searchable brain.**

They can:

- draft SOPs from recordings or bullet points
- generate training manuals
- update policies when things change
- answer internal questions ("How do we handle refunds?")
- store and organize your best prompts and playbooks

As you grow, this becomes essential for consistency and training.

CATEGORY 10 — Industry-Specific AI Tools

These are tools built around your exact type of business.

Examples

- **Construction:** Buildxact AI, Autodesk AI (estimating, plans, project management).
- **Medical/Healthcare:** DrChrono AI, Tebra AI (charting, patient communication).

- **Restaurants:** Toast AI, Otter AI (orders, delivery, menu optimization).
- **Retail/Ecommerce:** Shopify Magic, Klaviyo AI (product recommendations, email flows).
- **Legal:** Practice-management tools with AI summarization and drafting.

These tools are often **incredibly powerful** but should come **after** you've nailed the basics:

- AI writing
- customer response
- CRM
- scheduling
- automation

SECTION 3 — Building *Your* Personalized SMB AI Tech Stack

Your goal is NOT:

- "Adopt all 10 categories right now."

Your goal IS:

Pick 1–2 categories that solve your biggest problems.
Choose 1 tool in each.
Expand over 12 months.

A strong, lean AI stack for most SMBs looks like:

- **1 AI writing tool** (ChatGPT or Claude)

- **1 AI communication bot** (Tidio, Intercom, Air.ai, etc.)
- **1 CRM with automation** (HubSpot, GoHighLevel, etc.)
- **1 marketing/design tool** (Canva AI, Jasper, etc.)
- **1 automation engine** (Zapier or Make)

Everything else is optional and can be added later.

This aligns with your **AI Opportunity Matrix**:

- Start with **high-impact, low-complexity** categories.
- Save high-impact, high-complexity tools for later phases.

SECTION 4 — The Default SMB AI Tech Stack Blueprint

Here is the baseline architecture I implement for many clients.

Core Brain

- **ChatGPT or Claude** → Writing, planning, analysis, templates
- **Zapier** → Connects your tools and automates movement

Customer-Facing Layer

- **Tidio or Air.ai** → AI receptionist for chat/voice
- **HubSpot or GoHighLevel** → CRM + sales automation

Marketing Layer

- **Canva AI** → Graphics, social posts, simple design
- **Jasper or equivalent** → Marketing copy, campaigns

Operations Layer

- **Calendly** → Scheduling and reminders
- **QuickBooks AI** → Finance automation and visibility

This exact stack, tuned per business, works for **90% of small and medium service businesses**.

SECTION 5 — The Biggest Mistakes to Avoid

Avoid these, and you're already ahead of most of the market:

X Buying too many tools at once

X Choosing the most complex tool because it "does everything"

X Trying to automate your entire business in one quarter

X Failing to designate an internal *AI Champion*

X Not building a prompt library for repeatable tasks

X Not connecting tools with Zapier/Make

X Expecting AI to be perfect without human oversight

Think like a builder:
One brick at a time. But each brick placed intentionally.

SECTION 6 — A 30-Day Implementation Plan for Your AI Tech Stack

Here's how to build (or upgrade) your stack without chaos.

Week 1 — Choose Your 1–2 Highest-ROI Categories

- Use your **AI Opportunity Matrix**
- Pick the categories that fix your biggest bottlenecks (communication, leads, scheduling, etc.)

Week 2 — Choose ONE Tool Per Category

- Evaluate 3 tools max
- Pick the **simplest** that solves your problem
- Keep cost under **$50–$100/month** unless it clearly makes money

Week 3 — Train ONE AI Champion

- Choose one person who will own implementation
- Give them time to learn, test, and document
- Plug at least one workflow into each chosen tool

Week 4 — Automate ONE Workflow Per Tool and Measure

- Communication tool → one customer-facing workflow
- Writing tool → one content workflow
- CRM tool → one follow-up workflow
- Automation tool → one cross-app workflow

End of 30 days:

- Fewer manual tasks
- Shorter response times
- Cleaner communication
- Clearer visibility
- Real-world data on what works

SECTION 7 — Final Tech Stack Checklist

By the end of this chapter (or this quarter), aim to check these off:

✓ I identified 1–2 high-ROI workflows to improve
✓ I chose the right categories to support those workflows
✓ I selected **one tool** per category — no tool sprawl
✓ I connected my tools with Zapier/Make where needed
✓ I assigned an AI Champion on my team
✓ I launched at least one automated workflow per tool
✓ I measured time saved, leads captured, and errors reduced
✓ I documented what we built so it can be improved or scaled

You don't need a "perfect" tech stack.

You need a **simple, focused, AI-powered stack** that makes your business faster, leaner, and easier to run.

That's what you have the blueprint for now.

A Moment for Reflection

You've made incredible progress to get to this point.

But I want to be sure that you are not feeling overwhelmed. There is a lot to digest in this book. Much of it may be new to you. But I designed it to be a guide not a burden. I want to remind you that progress is much greater than perfection.

One workflow is better than ten tools. If you've implemented one to three of the ideas in this book so far, you are ahead of your competition by a long way.

You see, most business owners and people in general have the best intentions but are easily distracted by outside forces.

You see, most business owners—and people in general—have the best intentions, but are constantly being pulled in a hundred different directions.

A new tool launches.
A competitor tries something flashy.
A video pops up promising "10X growth overnight."
An article tells you that if you're not doing everything, you're already behind.

And before you know it, your focus is gone.

You start second-guessing what you've already built.
You start chasing instead of executing.
You start adding instead of refining.

And that's where momentum quietly dies.

Not because you aren't capable.
Not because you don't care.
But because your attention has been scattered.

Let me give you something simple to hold onto:

Distraction is the real competition.

Not AI.
Not other businesses.
Not technology.

The businesses that win are not the ones doing the most.

They are the ones doing **the right things consistently**, without getting pulled off course every week.

That's why I keep bringing you back to simplicity.

If you've implemented one workflow from this book—just one—you've already done something most business owners won't do.

You took action.

You didn't just read.
You didn't just think about it.
You didn't just talk about it.

You did something.

And that matters more than almost anything else.

Because execution compounds.

One workflow turns into two.
Two turn into a system.
A system turns into a competitive advantage.

And suddenly, without realizing it, your business starts to feel different.

More controlled.
More predictable.
Less chaotic.

Not perfect—but moving forward.

And that's the goal.

Not perfection.

Progress.

So if you're feeling that pull right now—the urge to jump ahead, to try everything, to overhaul everything at once—I want you to resist it.

Come back to this:

What is the one thing I can implement this week that will actually make my business better?

Not ten things.

Not someday.

This week.

Do that.

Then do it again next week.

And the week after that.

Because the businesses that win in this next era of AI won't be the ones that tried everything.

They'll be the ones that **stayed focused long enough to make a few things work really well.**

Stay in that lane.

You're closer than you think.

CHAPTER 7: AI for Marketing & Lead Generation

How Small Businesses Can Attract More Customers, Dominate Their Market, and Outperform Bigger Competitors Using AI

INTRODUCTION

Marketing Is No Longer Just About Creativity — It's About Systems, Speed, and Relentless Consistency

When I ask small business owners which part of their business frustrates them the most, the answer I hear more than any other is:

"Marketing. I never feel like I'm doing enough, and I never know if what I'm doing is working."

And they're not wrong.

Marketing today is:

- complicated
- competitive

- time-consuming
- constantly changing
- expensive if done wrong
- overwhelming if done manually

Most small businesses quietly suffer from a combination of:

1. Inconsistent posting
2. Weak or outdated branding
3. Slow response times
4. Not enough content
5. Not enough reviews
6. Poor lead follow-up
7. No system behind their marketing
8. Little to no data or tracking

And here's the part that stings:

You're not just competing with other small businesses anymore.

You're competing with:

- national franchises with big budgets
- global e-commerce companies
- local competitors using AI to create 100x more content than you
- agencies running AI-powered campaigns
- algorithms that reward daily activity

- customers who expect instant answers and personalized communication

Marketing is where AI changes everything.

Because AI doesn't just create more content.

It creates **consistent, on-brand, targeted content** that supports real business goals — and it does it:

- faster
- cheaper
- more consistently
- with better optimization
- without burning anyone out

This chapter gives you:

- the AI tools that matter
- the frameworks behind an AI-powered marketing engine
- the workflows you can copy
- the sequences you can automate
- the ROI you can realistically expect
- the mistakes to avoid

By the time you finish this chapter, marketing stops being chaos and starts being a **machine**.

Let's build it.

SECTION 1 — Why Marketing Is the #1 AI Use Case for Small Businesses

If AI disappeared tomorrow, the first place small businesses would feel deep pain is **marketing**.

Here's why.

1. Marketing eats more time than almost anything else

Owners and staff spend hours:

- writing social posts
- designing graphics
- answering DMs and emails
- updating websites
- writing newsletters and promos
- brainstorming ideas
- managing reviews

AI reduces that time by **70–90%** when used correctly.

2. Marketing needs creativity — but creativity can be systemized

Marketing seems like "magic," but at scale it's just:

- ideas
- formats
- angles
- hooks
- offers
- consistency

AI can generate:

- content ideas
- captions and scripts
- email subject lines
- Call to Action (CTA) and offers
- variations to test

With the right prompts and brand rules, AI can outproduce most small teams **by an order of magnitude.**

3. Marketing demands consistency — and AI never gets tired

Algorithms reward:

- consistent posting
- frequent engagement
- steady communication

Humans:

- get busy
- get tired
- forget
- burn out

AI doesn't.

4. Marketing demands speed — AI responds instantly

Leads don't wait.

If you respond in hours or days, your competitor who responds in **seconds** wins.

AI:

- replies instantly
- gathers key info
- books appointments
- sends follow-up messages

That response speed alone can change your lead flow.

5. Marketing requires data — AI analyzes it without whining

Most small businesses:

- never look at their analytics
- don't track conversions seriously
- guess what's working

AI can:

- summarize performance
- highlight best and worst content
- suggest what to do more of
- surface patterns you'd never see manually

This makes you more dangerous than competitors still "posting and praying."

SECTION 2 — The AI Marketing Stack: The Four Layers That Drive Lead Generation

Every effective marketing system can be broken into **four layers**.

AI supercharges each one.

LAYER 1 — Strategy (What You Say and to Whom)

This is the thinking layer.

AI helps you:

- define your ideal customer
- study competitors and identify gaps
- research keywords and search intent
- uncover emotional triggers and objections
- define your positioning and differentiators
- outline offers and lead magnets

You'll use AI as your **strategic thinking partner**, not just a content machine.

LAYER 2 — Content Creation (What You Produce)

This is the "stuff people see" layer.

AI can:

- write blog posts and articles
- draft emails and newsletters
- script videos and Reels
- create captions and hooks
- write landing page copy

- generate product descriptions
- draft ad copy

AI becomes your:

- copywriter
- scriptwriter
- brainstorming partner
- first-draft machine

You still control the direction. AI does the heavy lifting.

Research is a thousand times easier. You must edit, rewrite, and review your content so this is not a get out of work pass. But, partnering with AI will put you very far ahead of your competitors.

LAYER 3 — Distribution (Where It Goes and How Often)

This is where most businesses fail.

They have ideas.
They even create some content.
They just don't distribute it consistently.

AI-powered tools can:

- schedule posts across platforms
- send newsletters and autoresponders
- manage SMS campaigns
- update your Google Business Profile
- repurpose content to multiple channels

This allows you to show up **everywhere your customer lives**, without being chained to your phone.

LAYER 4 — Optimization (How It Gets Better Over Time)

This is the layer that turns marketing into a compounding asset.

AI helps you:

- analyze performance automatically
- A/B test subject lines and hooks
- identify which posts drive leads vs. vanity likes
- make data-backed content decisions
- refine your messaging and offers

When these four layers work together, your marketing stops being random and becomes an **AI-augmented system**.

SECTION 3 — The 10 Most Powerful AI Marketing Tools for SMBs

You don't need all of these, but you should recognize them and understand what role they play.

Think of this as your **AI Marketing Toolbox**.

Tool #1 — ChatGPT / Claude (Core Writing & Strategy Engine)

Use for:

- content calendars

- blogs and articles
- ad drafts
- email campaigns
- landing page copy
- scripts for Reels, podcasts, YouTube
- refining your brand voice
- keyword and topic research

These are your **central brain tools** for written and strategic work.

Tool #2 — Canva AI (Design & Visuals Engine)

Use for:

- social media graphics
- post templates
- flyers, postcards, banners
- simple brochures and one-sheets
- product image layouts
- event promos

Canva AI lets you create professional visuals **without** a designer on staff.

Tool #3 — Jasper (Long-Form Marketing Engine)

Use for:

- SEO blog posts
- long-form sales pages

- detailed email sequences
- product launch campaigns
- nurture sequences

Especially useful if you want long-form written content at scale.

Tool #4 — Perplexity (Research Engine)

Use for:

- competitor analysis
- researching topics your audience cares about
- finding stats and market trends
- summarizing complex topics into customer-friendly language

Think of this as "research on demand" so your content isn't just fluffy — it's informed.

Tool #5 — Vidyo.ai / OpusClip (Short-Form Video Engine)

Use for:

- turning long videos (webinars, talks, podcasts) into short Reels/TikToks
- automatic captioning
- finding the best snippets automatically
- generating viral-style cuts

Video is non-negotiable. These tools make it feasible.

Tool #6 — Reclaim AI (Scheduling & Focus Engine)

Use for:

- blocking time for content batching
- protecting time for marketing tasks
- ensuring you actually execute your marketing plan

Marketing doesn't work if it never gets your calendar time. Reclaim helps that.

Tool #7 — HubSpot AI or GoHighLevel (CRM + Follow-Up Engine)

Use for:

- automated email and SMS sequences
- lead tagging and segmentation
- pipeline tracking
- triggers based on behavior (clicks, opens, visits)
- reminders to call or follow up

This is where lead generation turns into **actual sales.**

Tool #8 — Surfer SEO (Search Optimization Engine)

Use for:

- SEO outlines
- keyword optimization
- content scoring vs top-ranking articles
- identifying what you need to add to compete

Perfect if SEO is part of your long-term traffic strategy.

Tool #9 — Tidio AI (Instant Response & Capture Engine)

Use for:

- website chat
- instant answers to FAQs
- capturing email/phone
- booking appointments

This tool stops your website from acting like a static brochure and turns it into a **24/7 salesperson**.

Tool #10 — Zapier (Distribution & Automation Glue)

Use to connect:

- website → CRM
- CRM → email/SMS
- forms → spreadsheets → notifications
- booking tools → reminders → follow-up sequences

Zapier makes sure that when someone engages with your marketing, the right things happen automatically.

You don't need all 10 to win.

For many small businesses, a **starting AI marketing stack** looks like:

- ChatGPT or Claude
- Canva AI
- Tidio

- HubSpot or GoHighLevel
- Zapier

Everything else can be layered in later.

SECTION 4 — The AI Marketing Operating System (AI-MOS)

Your Complete SMB Blueprint

This is the **system** I implement for clients.

It's your **AI Marketing Operating System (AI-MOS)** — and yes, this section alone is worth real consulting money.

We're going to combine:

- brand voice
- content planning
- content creation
- distribution
- follow-up
- reviews
- ads
- analytics

Into one coherent engine.

STEP 1 — Create Your AI-Enhanced Brand Voice

AI will eventually write 70–90% of your content.
If you don't define your brand voice, you'll get generic mush.

You need a **2-page Brand Bible** that includes:

- tone (friendly, expert, casual, edgy, luxury, etc.)
- vocabulary (what you say / don't say)
- ideal customer avatar
- top pains and desires
- your positioning ("We're the X who does Y for Z.")
- your values and promises
- examples of "good" content in your voice

You then **feed this into AI** and use prompts like:

"Write this in our brand voice (see Brand Bible below)."

This is how you keep AI on-brand.

STEP 2 — Build a 12-Month Content Strategy Using AI

Ask your AI writing tool to help you create:

- **12 monthly themes** (aligned with seasons, promotions, events)
- **52 weekly topics** (educational, authority-building, local/community)
- **365 short-form content ideas** (tips, FAQs, myths, stories, quick wins)

You don't need to use them all, but you'll never say "I don't know what to post" again.

STEP 3 — Turn Strategy into a Content Calendar

With AI, build a calendar that includes:

- **Daily** social posts (or 3–5x/week if that's realistic)
- **Weekly** short-form videos (Reels, Shorts, TikTok)
- **Bi-weekly** email newsletters
- **Monthly** blog posts or long-form content
- **Quarterly** campaigns or promotions
- **Regular** review requests and testimonials

AI can output this in a structured format (table/spreadsheet) your team can use.

This shifts you from **random posting** to a **planned, strategic schedule.**

STEP 4 — The AI Content Assembly Line

Here's your new production model:

PLAN → WRITE → DESIGN → SCHEDULE → OPTIMIZE

1. **PLAN** — AI helps you choose which ideas to publish this week.
2. **WRITE** — AI drafts posts, emails, and scripts.
3. **DESIGN** — Canva AI creates graphics and layouts.
4. **SCHEDULE** — Use scheduling tools or Zapier-driven flows.
5. **OPTIMIZE** — AI reviews performance and suggests tweaks.

Done right, **1–2 hours of focused work per week** can produce weeks of content.

STEP 5 — Build AI Follow-Up Engines

Lead generation without follow-up is wasted money.

AI allows you to:

- send instant email replies when someone fills a form
- send instant SMS replies when someone texts or opts in
- reply immediately on website chat
- build **nurture sequences** that warm leads over days/weeks
- score leads based on behavior (opens, clicks, visits)
- trigger booking links automatically

This alone can increase revenue **20–40%** without more ad spend.

STEP 6 — Build AI Review & Reputation Systems

Reviews are marketing rocket fuel.

AI can help you:

- draft review request messages
- send them automatically after visits or purchases
- respond to reviews in your voice
- escalate negative reviews internally
- pull in reviews to highlight on your site and social

Instead of manually begging for reviews, you have a **review machine** running in the background.

STEP 7 — Use AI to Power Your Ad Engines (Facebook, Google, Local)

AI won't replace smart ad strategy, but it makes executing that strategy dramatically easier.

AI can:

- write multiple ad copy variations
- generate creative ideas
- help brainstorm targeting options
- analyze which ads are working
- suggest angles to test (price, speed, quality, social proof, etc.)

You (or your agency) can iterate much faster, test more, and waste less money.

STEP 8 — Build an AI-Enhanced Measurement Dashboard

Most owners never look at their metrics because they don't know where to start.

AI can:

- summarize your marketing performance weekly
- pull in data from multiple tools (via exports or summaries)
- highlight what's up, what's down, and what changed
- translate "nerd data" into "owner language"

Ask AI:

"Here are my last 7 days of marketing stats. Summarize what happened, what's working, and what I should do next."

Now you're managing like an owner, not guessing like a gambler.

SECTION 5 — AI-Driven Marketing Workflows (By Industry)

Here are some plug-and-play examples.

Medical Spas

AI handles:

- before/after post captions
- educational posts ("What is microneedling?")
- appointment reminders and follow-ups
- care instructions after treatments
- seasonal promos (Valentine's Day, Summer Glow, Holiday Skin)
- review requests and responses
- Q&A content for website and social

Result: more bookings, better education, stronger retention.

Restaurants

AI handles:

- daily specials posts
- menu item descriptions
- event and catering promos
- holiday menus and campaigns
- chef spotlights and behind-the-scenes posts

- reservation FAQs and hours
- review responses

Result: more foot traffic, higher check size, stronger local brand.

Home Services (HVAC, Plumbing, Roofing, Landscaping)

AI handles:

- service explanations (plain-English breakdowns)
- seasonal reminders (AC tune-ups, winter prep, gutter cleaning)
- estimate email templates
- follow-up on unsold estimates
- tips and homeowner guides
- Google Business Profile posts
- safety and maintenance content

Result: more booked jobs, higher trust, repeat customers.

Real Estate Teams

AI handles:

- listing descriptions
- neighborhood and local business spotlights
- weekly market update emails
- buyer and seller nurture sequences
- social content showcasing closings and stories
- lead prequalification scripts

- open house follow-up

Result: more appointments, higher engagement, better perceived authority.

E-commerce & Retail

AI handles:

- product descriptions
- abandoned cart emails
- cross-sell and upsell flows
- seasonal campaigns
- UGC prompts ("Share a photo and tag us!")
- email campaigns for drops and launches

Result: higher conversion rates, more repeat purchases, stronger brand.

SECTION 6 — Why AI Marketing Works: Consumer Psychology

AI-driven marketing works because it lines up perfectly with how modern customers behave.

Customers today expect:

- **instant answers** → AI-powered chat and autoresponders
- **personalized content** → segmentation + AI copy
- **frequent updates** → scheduled content and campaigns

- **consistent branding** → AI trained on your Brand Bible
- **high-quality visuals** → Canva AI, video tools
- **fast follow-up** → CRM + AI sequences
- **proof and social validation** → reviews, testimonials, stories

AI doesn't change what people want.
It just finally gives small businesses the **capacity** to deliver it.

SECTION 7 — AI Marketing ROI: What You Can Realistically Expect

When implemented correctly, small businesses typically see:

- **30–50% increase in leads**
- **2x–4x increase in content output**
- **40–70% reduction** in manual marketing labor
- **50% faster lead response times**
- **20–40% increase** in engagement and inquiries
- **improved Google Business Profile ranking** from consistency
- **more reviews**, faster

These results usually show up over **60–120 days**, not overnight.

But they are very real.

SECTION 8 — The Biggest Marketing Mistakes to Avoid with AI

Let's keep you out of the ditches.

1. **Letting AI produce generic content**
    - Fix: Train it with your Brand Bible and examples.
2. **Posting without strategy**
    - Fix: Start with business goals → offers → then content.
3. **Trying to automate everything at once**
    - Fix: Use the **One-Workflow Rule™** — pick one workflow at a time.
4. **Using too many tools**
    - Fix: Start with 3–5 core tools and master them.
5. **Expecting AI to fix a bad offer or bad product**
    - AI will amplify your strategy — good or bad.
6. **Removing humans from the loop entirely**
    - AI should assist.
    - Humans still approve, refine, and direct.

SECTION 9 — The 30-Day AI Marketing Launch Plan

Here's a realistic, high-leverage 30-day rollout.

Days 1–3 — Build Your Brand Voice

- Create your 2-page Brand Bible
- Feed it to your AI writing tool

- Test a few posts or emails to tune the voice

Days 4–6 — Build a 12-Month Content Framework with AI

- Map 12 monthly themes
- Map 52 weekly topics
- Generate a bank of daily ideas

Days 7–14 — Produce and Schedule 30 Days of Content

- Use AI to draft posts, emails, simple scripts
- Use Canva AI for visuals
- Schedule everything for the next 30 days

Days 15–20 — Build Your Follow-Up Engines

- Create AI-assisted email/SMS sequences for new leads
- Connect website/chat → CRM → sequences

Days 21–25 — Build Your Review & Reputation System

- Create AI-generated review request messages
- Set triggers (after visits, purchases, completed jobs)
- Draft standard AI-assisted responses for reviews

Days 26–30 — Build Your Analytics Briefing

- Decide what metrics matter (leads, bookings, opens, clicks)
- Export or connect data
- Ask AI for weekly performance summaries

By Day 30, you don't just "use AI in marketing."

You have a **marketing engine** that runs with AI.

SECTION 10 — The Future of AI Marketing (And Why You Need to Move Now)

AI is pushing marketing from:

- **guesswork → data-driven systems**
- **random posting → planned content engines**
- **manual writing → assisted, scalable content**
- **inconsistent messaging → brand-consistent communication**
- **slow follow-up → instant engagement**
- **expensive agencies → affordable, in-house capability**

Local and small businesses that adopt AI-powered marketing now will:

- show up more
- respond faster
- look more professional
- close more leads
- build deeper trust

The ones who don't will tell themselves:

"We just can't compete anymore."

The truth is, they could have.

They just didn't build the system.

SUMMARY: What You Built in This Chapter

By completing this chapter, you now have:

✓ A clear understanding of why marketing is AI's #1 use case
✓ A four-layer AI marketing stack (Strategy, Creation, Distribution, Optimization)
✓ A practical AI Marketing Toolbox with 10 key tools
✓ An **AI Marketing Operating System (AI-MOS)** blueprint
✓ A Brand Voice "Bible" concept to keep AI on-brand
✓ A content assembly line you can use every month
✓ AI-powered follow-up and review systems
✓ Industry-specific workflow examples you can copy
✓ A 30-day rollout plan to implement all of this
✓ A realistic sense of the ROI you can expect

Next, we move deeper into **AI for Sales & Follow-Up** — where all this marketing momentum turns into signed contracts, booked jobs, and real revenue.

Because attention alone doesn't pay the bills.

Conversion does.

CHAPTER 8: AI for Sales & Follow-Up

How to Stop Losing Leads, Close More Deals, and Build a Relentless Revenue Engine

INTRODUCTION

You're Not Losing Because Your Offer Is Bad — You're Losing Because Your Follow-Up Is Weak

Most small business owners don't have a sales problem.

They have a **follow-up problem**.

Here's the pattern I see over and over:

- leads come in
- someone says "I'll follow up later"
- the day gets busy
- sticky notes pile up
- emails sit half-written
- calls never get returned
- quotes go unsent
- proposals go un-followed

And at the end of the month, the owner says:

"We just need more leads."

No, you don't.

You need to stop wasting the ones you already have.

Multiple studies show that roughly half of leads never get a meaningful follow-up, and around 80% of marketing leads never convert—most often because they're never properly nurtured.

- MarketingSherpa reports that 79% of marketing leads never convert into sales, primarily due to lack of lead nurturing. Most businesses stop after 1–2 attempts.
- A Martal Group 2026 lead generation benchmark cites the same figure: about 79% of leads never convert into sales, often because they lack proper nurturing or qualification.
- A SalesMasters article referencing separate research notes that 23% of companies never responded to inbound leads at all, and another 24% took more than 24 hours, effectively "losing" nearly half of inbound leads on basic follow-up speed.
- ZoomInfo (using a compilation of industry studies) reports that "the average sales rep only makes 2 attempts to reach a prospect," and 44% of salespeople give up after one follow-up.
- An Invesp sales follow-up piece cites very similar numbers: "48% of sales teams never even attempt a

follow-up after an initial interaction," and only 12% make three or more follow-ups.

- EmailToolTester's 2025 research on touchpoints shows warm inbound leads require 5–12 touches on average, while cold prospects can require 20–50 touches.
- Flowlu's 2026 sales statistics report says "80% of sales require 5 to 12 follow-up attempts," and 92% of salespeople stop following up after four or fewer attempts
- the business that responds fastest usually wins
- A LinkedIn summary of the MIT Lead Response Management study shows that effectiveness drops from a 100% baseline at under 1 minute to about 10% effectiveness if you wait 24+ hours, implying a massive penalty for slow responders.

This is where AI becomes a weapon.

Because AI doesn't just help you get more leads (marketing).

It helps you **convert** them:

- faster
- more consistently
- more intelligently
- with less manual effort

In this chapter, you'll learn how to use AI to:

- respond to leads instantly
- follow up automatically without sounding robotic
- send quotes and proposals faster
- keep deals moving without you chasing every thread
- build simple, aggressive, ethical follow-up systems
- make your sales process feel like a concierge, not chaos

You'll build an AI-assisted sales engine that doesn't forget, doesn't get tired, and doesn't stop after the first "no response."

Let's get into it.

SECTION 1 — Why Small Businesses Bleed Revenue in Sales & Follow-Up

Before we talk about AI, we need to be blunt about where money is leaking.

Most small businesses lose deals because they:

1. **Respond too slowly**
    - The lead fills out a form at 8:47 p.m.
    - They hear back… "sometime tomorrow"
    - A competitor responded in 30 seconds with a text and booking link
    - Game over
2. **Stop following up too early**
    - 1 email

- maybe 1 call
- then silence

3. **Have no standardized process**
    - every salesperson does their own thing
    - no scripts
    - no structured offers
    - no defined touchpoints
4. **Spend time on the wrong leads**
    - chasing unqualified people
    - ignoring high-intent prospects
5. **Let quotes and proposals die quietly**
    - a quote goes out
    - nobody checks in
    - no reminders
    - no "What questions can I answer?"
    - no "Do you want to move forward?"
6. **Forget that buyers are busy, not rude**
    - the prospect didn't ghost you
    - they got busy
    - the business that stays top-of-mind wins

You don't fix these with more hustle.

You fix them with **systems**.

AI gives you the ability to build those systems without hiring a full-time sales admin.

SECTION 2 — The Four Pillars of an AI-Enhanced Sales System

A good AI sales system has four main pillars:

1. **Speed-to-Lead**
2. **Structured Follow-Up**
3. **Smart Prioritization**
4. **Owner Visibility & Accountability**

AI plugs into each one.

Pillar 1 — Speed-to-Lead

First to respond often wins. Inside Sales reports conversion rates are more than 8x higher when leads are contacted in the first 5 minutes vs 5 minutes–24 hours.

AI lets you:

- reply to inquiries in seconds
- acknowledge the lead
- ask a few qualifying questions
- offer a next step (booking link, call, estimate)

This alone can change your close rate.

Pillar 2 — Structured Follow-Up

No more "winging it."

AI helps you:

- design follow-up sequences (email + SMS + calls)
- write all the messages in your tone
- adjust content based on how they respond
- keep going for days or weeks — not just once

Pillar 3 — Smart Prioritization

Not all leads are equal.

AI + your CRM can:

- score leads based on behavior (clicks, opens, visits, time spent)
- flag "hot" leads
- suggest who to call today
- summarize key info before you get on the phone

This lets humans focus where they matter most.

Pillar 4 — Owner Visibility & Accountability

You need to know:

- how many leads came in
- who followed up
- how fast
- how many converted

- where deals are stuck

AI can summarize this for you so you're not digging through spreadsheets at midnight.

SECTION 3 — The AI Sales Tech Stack: Tools That Actually Move the Needle

You don't need fancy enterprise systems.

For most small businesses, a lean AI sales stack looks like:

- **AI Writing/Reasoning Tool:** ChatGPT / Claude
- **CRM with Automation:** HubSpot / GoHighLevel / Close
- **Instant Response Tool:** Tidio / Air.ai / website chat
- **Automation Engine:** Zapier / Make
- **Scheduling Tool:** Calendly / similar

Let's break down how they work together.

1. AI Writing/Reasoning (ChatGPT / Claude)

Use this to:

- write sales scripts and objection handlers
- draft outreach messages
- rewrite follow-up emails in your tone
- summarize long email threads into "What's going on here?"
- generate proposal outlines

This is your "sales copy on demand" engine.

2. CRM with AI & Automation (HubSpot / GoHighLevel / Close)

Use this to:

- capture leads automatically
- trigger sequences (email, SMS, tasks)
- track stages (New Lead → Contacted → Qualified → Proposal Sent → Closed Won/Lost)
- log activities without manual typing
- send reminders to your salespeople

This is the **heart** of your AI sales engine.

3. Instant Response Tools (Tidio / Air.ai / Chat Widgets)

Use these to:

- greet website visitors
- answer basic questions
- capture contact information
- route qualified leads into your CRM
- offer booking options immediately

Think of this as your **AI front desk**.

4. Automation Engine (Zapier / Make)

Use this to glue everything together:

- website form → CRM → SMS + email → internal notification
- new lead in CRM → apply tag → add to follow-up sequence
- appointment booked → confirmation email + SMS → add note
- quote sent → delay → follow-up message if no response

5. Scheduling Tool (Calendly / Acuity Scheduling/ Doodle/ Setmore/ SimplyBook.me/ YouCanBookMe/ Koalendar/ TidyCal/ Trafft)

Use this as the **default next step**:

- "Pick a time here to talk."
- "Choose a consultation slot that works for you."
- "Book your estimate in 30 seconds here."

When leads can book themselves, you close more without chasing.

SECTION 4 — Speed-to-Lead: Building Your AI-Powered First Response

Your first response doesn't have to close the deal.

It just has to:

1. **Be fast**
2. **Acknowledge the inquiry**
3. **Gather key info**

4. **Offer a clear next step**

AI can handle this perfectly.

Example: Website Form → Instant AI Reply

Workflow:

1. Lead fills out form on website.
2. Zapier sends data to CRM and triggers autoresponder.
3. AI-crafted email/text goes out instantly.

Sample AI-Assisted First Email:

Subject: Got your request — let's get you taken care of

Hi {{FirstName}},

Thanks for reaching out about {{service/interest}}. We've received your request and you're on our radar.

To make sure we give you the most accurate quote/recommendation, could you quickly tell us:

1. {{Key qualifier #1}}
2. {{Key qualifier #2}}

If it's easier, you can skip all that and pick a time to talk here: {{CalendlyLink}}

Looking forward to helping you,
{{YourName}}
{{BusinessName}}

You write the first version with AI, then refine it until it feels like **you**. After that, it runs on autopilot.

SECTION 5 — The AI-Assisted Follow-Up Ladder™

Most sales happen after multiple touches, not the first.

Here's a simple follow-up ladder that AI can help write and your CRM can automate:

Day 0: Instant confirmation + basic info request
Day 1–2: "Here's a bit more about how we can help" (value email)
Day 3–4: "Here are common questions people ask"
Day 5–7: "Are you still interested in X?"
Day 10–14: "Just checking in — want to move forward or pause for now?"
Day 21–30: "We haven't heard back — here's a helpful resource or limited-time incentive."

AI's job:

- write each email/text in your brand voice
- generate variations for testing
- adapt messages to different services or segments

Your job:

- approve
- tweak
- monitor

Sample AI-Assisted Follow-Up Sequence (Email + SMS)

Email #1 (Day 1): Value & Reassurance

Subject: A quick look at how we help clients like you

Hi {{FirstName}},

Thanks again for reaching out about {{service}}. I wanted to give you a clearer picture of what working with us actually looks like.

Here's what most of our clients want:

- {{Outcome 1}}
- {{Outcome 2}}
- {{Outcome 3}}

And here's how we help them get there:

- {{Short bullet process}}

If you'd like, you can grab a quick spot on my calendar here: {{CalendlyLink}}

Talk soon,
{{YourName}}

SMS #1 (Day 2): Light Touch

Hi {{FirstName}}, this is {{YourName}} from {{BusinessName}}. Just wanted to see if you had any questions about {{service}} or if you'd like a quick call to go over options.

You can grab a time here: {{CalendlyLink}}

AI makes writing these fast. Your CRM makes sending them automatic.

SECTION 6 — AI for Objections, Proposals, and Closing Support

AI can't close deals **for** you.

But it can make you dramatically more prepared.

Handling Objections with AI

Common objections:

- "It's too expensive."
- "We need to think about it."
- "We're talking to other providers."
- "Now's not the right time."

You can feed transcripts or notes into AI and ask:

"These are the common objections we hear. Help me craft respectful, confident responses in my voice that focus on value, not pressure."

You'll walk away with:

- written responses
- scripts
- email templates

You then practice and adapt them.

Using AI to Draft Proposals Faster

Instead of staring at a blank screen, you:

1. Give AI a simple outline:
    - client type
    - services requested
    - scope
    - price range
2. Ask it to draft:
    - a structured proposal
    - key benefits
    - a simple summary page
3. You edit, personalize, and send.

Result: proposals in **15–30 minutes** instead of half a day.

AI for "Stuck Deals"

If you've had a proposal out for weeks with no response, AI can help you write:

- a "soft nudge" email
- a "close the loop" email
- a "last chance before we release your spot" email

These are often uncomfortable to write.
AI makes it less emotional and more systematic.

SECTION 7 — Industry-Specific AI Sales Workflows

Let's make this concrete across a few types of businesses.

Home Services (HVAC, Plumbing, Roofing, Landscaping)

Workflow:

1. Website lead form completed
2. Instant AI response (email + SMS)
3. AI asks for photos / details if needed
4. CRM sequences follow-up if no booking
5. AI drafts estimate email once scope is known
6. Follow-up sequence if estimate not accepted in 5–7 days

Results:

- less time chasing
- faster quotes
- more closed jobs from the same number of leads

Med Spas & Clinics

Workflow:

1. A potential customer inquires about a treatment
2. AI sends information about the treatment + before/after expectations
3. AI offers consultation booking link
4. Follow-up sequence for those who don't book
5. Post-consultation: AI drafts recap email + next steps

Real Estate

Workflow:

1. Buyer submits info on a listing
2. AI sends property details + 2–3 similar listings
3. AI offers a time for a call or showing
4. AI drafts follow-up emails with market updates
5. Lead scoring highlights the most engaged buyers

B2B Services (Consultants, Agencies, Accountants, Lawyers)

Workflow:

1. Prospect downloads a lead magnet or fills contact form
2. AI sends welcome email + calendar link
3. CRM starts a nurture sequence with helpful content
4. AI helps summarize discovery calls into "Deal Notes"
5. AI drafts proposals and engagement letters

SECTION 8 — Metrics That Matter in an AI-Driven Sales System

AI doesn't remove the need for metrics.
It makes them easier to see.

You should regularly track:

- **Leads In**
- **Speed-to-Lead** (time from inquiry to first response)
- **Contact Rate** (how many leads you actually talk to)
- **Booked Appointments**

- **Show Rate** (for calls/appointments)
- **Close Rate**
- **Average Deal Value**
- **Sales Cycle Length** (time from first contact to close)

You can feed snapshots of CRM exports into AI and ask:

"Summarize my sales performance this month and identify where deals are dropping off."

It will help you see whether:

- you have a top-of-funnel problem (not enough leads)
- a follow-up problem
- a close-rate problem
- a pricing/offer problem

Then you fix the right thing.

SECTION 9 — Common Mistakes When Using AI in Sales

Avoid these and you're ahead of most of the field:

1. **Letting AI speak like a robot**
    - Fix: Always train on your Brand Bible and edit your initial templates.
2. **Over-automating and under-humanizing**
    - Fix: Use AI for first responses and light touches, but keep humans involved in serious conversations and closing.

3. **Not aligning sales messages with marketing promises**
    - Fix: Use the same AI brand prompts for both marketing and sales so they feel unified.
4. **Ignoring reply signals**
    - If someone clicks, opens repeatedly, or replies but isn't ready, don't drop them.
    - Use AI to suggest the next best message.
5. **Using AI as a crutch instead of a coach**
    - AI should help you get better at sales conversations, not replace them.

SECTION 10 — Your 30-Day AI Sales & Follow-Up Launch Plan

Here's a simple rollout plan to match the rest of your 30-day roadmaps.

Days 1–5 — Map Your Sales Process

- How do leads come in today?
- What happens next?
- Where do they drop off?
- Where do you currently rely on memory, sticky notes, or chaos?

Document the REAL process, not the ideal one.

Days 6–10 — Build Your Speed-to-Lead System

- Choose your instant response method (chat, auto-email, SMS).
- Write your first-response templates with AI.
- Connect your website forms to your CRM and autoresponders.

Goal: every lead gets a reply in under 1–2 minutes, 24/7.

Days 11–15 — Build a Simple Follow-Up Ladder

- Use AI to write a 5–7 step follow-up sequence (email + optional SMS).
- Plug it into your CRM or automation tool.
- Apply it to ALL new leads of a certain type.

Days 16–20 — Upgrade Your Proposals & Objection Handling

- Use AI to create proposal templates.
- Build a library of objection responses.
- Practice using them so sales conversations improve.

Days 21–25 — Add Lead Scoring and Prioritization (If Your CRM Supports It)

- Define what "hot" activity looks like (opens, clicks, visits).
- Ask AI to help you create simple rules.
- Have the CRM surface "Today's Top Leads" daily.

Days 26–30 — Build Your Sales Performance Briefing

- Once a week, export or screenshot your sales stats.
- Ask AI to summarize them in plain language:
    - Where are we losing people?
    - What improved this week?
    - What should we test next?

By Day 30, you have:

- faster responses
- structured follow-up
- better proposals
- visibility into what's working

From there, you refine and expand — one workflow at a time.

MOVING FORWARD

AI Won't Close Deals for You but it Will Finally Let You Sell Like the Business You Could Be

AI is not here to replace your ability to build trust, read people, and make deals.

It's here to:

- remove the sloppiness
- remove the inconsistency
- remove the excuses
- remove the "I forgot to follow up"

It makes you:

- faster
- more organized
- more present
- more prepared

The businesses that win in the AI era won't necessarily have the slickest websites or the fanciest branding.

They'll be the ones who:

- respond first
- follow up consistently
- educate clearly
- remove friction
- stay top-of-mind

AI gives you the power to do all of that — with a small team, a reasonable budget, and a system that doesn't rely on your memory.

You just built the blueprint.

Next, we'll turn to **AI for Customer Service & Appointment Management**, where your sales promises are delivered — and your AI-powered customer experience keeps people coming back again and again.

CHAPTER 9: AI for Customer Service & Appointment Management

How to Deliver Faster, Friendlier, 24/7 Service Without Burning Out Your Team

INTRODUCTION

Customer Service Is Where You Win or Lose — AI Just Decides How Fast

Most small businesses don't lose customers because they're bad at what they do.

They lose customers because:

- nobody answered the phone
- an email sat in the inbox for two days
- a message on Facebook was never seen
- appointments got mixed up
- nobody confirmed or reminded them
- something slipped through the cracks

The actual **work** is usually good.
Everything around it? That's where chaos lives.

Modern customers don't just compare you to the business down the street.

They compare you to:

- Amazon's speed
- Uber's updates
- their bank's app
- their airline's text alerts

They're used to:

- instant answers
- real-time updates
- reminders
- self-service booking
- clear communication

If your business can't provide that kind of experience manually, you're not "bad," you're human.

AI is how you close that gap **without**:

- hiring a call center
- building a giant support team
- working 16-hour days on messages and scheduling

In this chapter, you'll learn how to use AI to:

- respond to customers instantly
- manage appointments with fewer mistakes
- reduce no-shows

- automate reminders and updates
- route issues to the right person
- keep customers feeling informed and cared for — at scale

Customer service is where your brand is proven or broken.

AI lets you show up like a **professional, modern company** even if you're a team of three.

Let's build that.

SECTION 1 — Why Customer Service and Scheduling Are Perfect for AI

AI is not equally useful everywhere.
Customer service and appointment management are **prime territory**.

Here's why.

1. It's Repetitive

Customers ask the same questions over and over:

- "What are your hours?"
- "Do you take my insurance?"
- "Do you service my area?"
- "What does this cost?"
- "Do you have availability this week?"

You don't need a human to answer those 500 times a month.

2. It's Time-Sensitive

Customers rarely sit around waiting:

- if they don't get a fast answer, they move on
- if they can't book quickly, they find someone else

AI can respond in **seconds** where humans respond in **hours** (or days).

3. It's Rule-Based

Appointment rules are logical:

- open hours
- staff availability
- service durations
- buffer times

AI + scheduling tools can follow those rules perfectly.

4. It's Emotionally Loaded

When customers are:

- confused
- worried
- frustrated
- on a deadline

...waiting makes the emotion worse.

AI reduces that wait, which reduces the negative emotion.

5. It's Expensive to Staff Manually

To provide “always on” support, you’d need:

- multiple front desk staff
- extended hours
- high payroll

AI lets a small team deliver large-company responsiveness.

SECTION 2 — The Four Pillars of AI-Enhanced Customer Service

A strong AI-powered customer service and scheduling system has four pillars:

1. **Instant Answers**
2. **Smart Routing & Escalation**
3. **Appointment Automation**
4. **Proactive Communication**

Let’s unpack each one.

Pillar 1 — Instant Answers

AI can:

- handle FAQs
- provide basic guidance
- share links and resources
- explain services in plain language

Channels:

- website chat

- SMS
- Facebook/Instagram DMs
- email autoresponders

Goal: no customer feels ignored.

Pillar 2 — Smart Routing & Escalation

AI should **not** handle everything.

It should:

- know when to escalate
- know who to escalate to
- collect key information first

Example:

- “Is this an emergency?”
- “Are you an existing customer?”
- “What’s your preferred contact method?”

Then it passes the conversation — with context — to the right person.

Pillar 3 — Appointment Automation

Scheduling is low creativity, high rules.

AI + scheduling tools can:

- let customers book online
- avoid double-booking
- handle time zones

- send confirmations
- manage reschedules and cancellations
- update your calendar automatically

Goal: fewer phone calls, fewer mistakes, fewer no-shows.

Pillar 4 — Proactive Communication

Customer experience soars when you communicate *before* they worry.

AI can help you:

- send reminders
- send prep instructions ("What to do before your appointment")
- send follow-ups ("How did it go?")
- send check-ins ("Need anything else?")

This makes your business feel **organized, thoughtful, and trustworthy**.

SECTION 3 — The AI Customer Service & Scheduling Stack

You don't need enterprise call center software.

Most small businesses can start with:

- **AI Chat/Response Tool:** Tidio, Intercom, Manychat, Drift
- **Voice AI (Optional):** Air.ai, Smith.ai (AI + human hybrid)

- **Scheduling Tools:** Calendly, TidyCal, or industry-specific systems
- **CRM / Practice Management / Booking System:** HubSpot, GoHighLevel, or your vertical tool
- **AI Writing Tool:** ChatGPT / Claude for scripts and templates
- **Automation Tool:** Zapier / Make to connect everything

We'll use these to build simple but powerful workflows.

SECTION 4 — Designing Your AI Customer Service Playbook

Before you plug in tools, you need a **playbook.**

This is where most businesses skip straight to "tools" and then get frustrated.

Your AI playbook should answer:

1. **What are the top 20 questions customers ask?**
2. **Which questions can AI answer confidently and safely?**
3. **Which questions must be escalated to a human?**
4. **What are your must-follow rules for appointments?**
5. **What tone should your customer service use?**

You already did a **Brand Bible** for marketing.
Here, you create a **Service & Scheduling Bible** for AI.

Include:

- tone: calm, helpful, local, non-pushy
- phrases to use (and avoid)
- escalation triggers ("If customer mentions X, send to human")
- your policies (cancellations, reschedules, deposits)
- scheduling rules (duration, buffers, staff)

You then feed this into your AI tools so they sound and behave like **your business**, not generic support.

SECTION 5 — AI Workflows: From Initial Inquiry to Completed Appointment

Let's walk through a full-flow example of how AI can support a typical customer journey.

We'll keep it universal so you can adapt it.

Step 1 — Initial Inquiry

Channel: website chat, contact form, SMS, social DM

AI Job:

- greet the customer
- ask 2–3 simple questions
- identify what they want
- collect name, email, phone

Example AI prompt logic:

"Ask for the customer's name, what they're looking for, and whether they prefer phone, text, or email."

Step 2 — Qualification (Light)

AI can ask:

- "Are you a new or existing customer?"
- "When are you hoping to book?"
- "Is this urgent?"

AI then:

- tags them in your CRM (new vs existing, urgency, service type)
- logs this info so your team sees it

Step 3 — Instant Info + Offer to Book

AI uses your knowledge base to provide:

- pricing ranges
- service details
- what to expect

Then it offers:

- "Would you like to book a time now?"
- [Booking link]

If they click, the scheduling tool handles the rest.

Step 4 — Confirmation & Reminders

Once an appointment is booked:

- your booking system sends a confirmation
- AI (or your system) sends reminders via email/SMS

- you can add "What to expect / how to prepare" messages

This reduces no-shows and preemptively answers questions.

Step 5 — Post-Appointment Follow-Up

After the appointment:

- AI sends a check-in message
- asks if they have questions
- optionally sends care instructions, next steps, or upsell offers
- sends a **review request** with a link

This is where service, marketing, and reputation all connect.

SECTION 6 — Industry-Specific AI Service & Scheduling Examples

Let's make this tangible.

Med Spa / Clinic

AI can:

- explain treatments (in plain English)
- answer "Is this right for me?" with disclaimers
- share before/after expectations (without making guarantees)
- book consultations
- send pre-appointment instructions ("No retinoids for 3 days...")

- send post-treatment care instructions
- request reviews

Result: fewer front desk calls, smoother patient flow, better experience.

Salon / Barber / Spa

AI can:

- answer pricing and availability questions
- help customers pick services
- offer specific providers if desired
- manage reschedules and cancellations
- send reminders
- send “It’s time for your next visit” messages

Result: fuller books, fewer gaps, lower no-show rate.

Home Services (HVAC, Plumbing, Electrical, Landscaping)

AI can:

- handle “Do you service my area?” and “What do you charge?”
- book estimate visits or service calls
- send pre-visit reminders (“Please secure pets,” etc.)
- send “tech on the way” messages (when integrated with field software)
- request a review after the job is complete

Result: less time on the phone, better customer communication.

Professional Services (Law, Accounting, Therapy, Coaching)

AI can:

- screen for basic fit ("We specialize in X, is that what you need?")
- share what documents they should bring
- manage consultation bookings
- send sensitive messages with careful tone (reviewed templates)

Result: more qualified leads on your calendar, less back-and-forth.

SECTION 7 — Guardrails: Where AI Should NOT Act Alone

This is critical.

AI can do a lot — but you must be clear where **humans must remain in control**.

Examples:

- medical advice and diagnosis
- legal opinions or instructions
- mental health and crisis situations
- financial advice about investments or taxes
- serious complaints or disputes

You should define:

- red-flag keywords ("suicidal," "illegal," "lawsuit," "emergency," etc.)
- topics AI should **only** respond to by saying:

"I'm so glad you reached out. This is something a member of our team needs to handle personally. I'm going to pass this to them right away."

You protect your customers, your team, and your liability by keeping humans in the loop where it matters most.

SECTION 8 — Metrics That Matter in AI-Enhanced Service & Scheduling

If you don't measure, you're guessing.

Key metrics:

- **First Response Time**
    - How long from customer inquiry to first response?
- **Resolution Time (For Simple Requests)**
    - How long until the customer gets what they needed?
- **Appointment Booking Rate**
    - Of inquiries, how many end in a scheduled appointment?
- **No-Show Rate**
    - How often do people miss appointments?

- **Customer Satisfaction**
    - Simple 1–5 rating after an interaction or visit
- **Review Volume and Rating**

You can feed snapshots of these into AI and ask:

"Summarize our customer service performance this month and tell me where we're improving and where we're failing."

You will see:

- where AI is working
- where humans need more training
- where processes need to be tightened

SECTION 9 — Common Mistakes When Implementing AI in Customer Service

Avoid these potholes.

1. **Making the AI pretend to be human**
    - Don't lie.
    - It's fine to say:

"I'm your virtual assistant here to help."

2. **Overcomplicating flows on day one**
    - Start with simple FAQs and basic booking.
    - Add complexity later.
3. **Not giving AI clear, written policies**

    - If policies live only in people's heads, AI will be inconsistent.
4. **Failing to train staff how to work with AI**
    - Staff should know what AI handles vs. what they own.
    - And how to step in gracefully when needed.
5. **Letting AI send sensitive messages unsupervised**
    - Some responses should be drafted by AI, approved by humans.
6. **Never reviewing conversations**
    - You must periodically audit AI interactions and improve.

SECTION 10 — 30-Day AI Customer Service & Scheduling Launch Plan

Here's your rollout roadmap.

Days 1–5 — Map Your Current Customer Journey

- How do customers contact you now?
- Where do they get stuck or ignored?
- How are appointments booked today?
- Where do mistakes happen?

Write it down honestly.

Days 6–10 — Build Your Service & Scheduling Bible

- Top 20 FAQs
- Policies (cancellations, reschedules, payments)
- Your desired tone
- Clear "escalate to human" rules
- Appointment rules (durations, buffers, staff)

Feed this into your AI tools.

Days 11–15 — Launch Basic AI Chat + FAQ Support

- Set up Tidio / Intercom / similar on your website
- Train it on your FAQs and policies
- Test in-house before turning it loose

Goal: AI handles simple questions reliably.

Days 16–20 — Connect Scheduling

- Integrate Calendly or your booking system
- Let AI offer booking links at the right moments
- Add automated confirmations and reminders

Goal: customers can book themselves without calling.

Days 21–25 — Build Simple Follow-Up & Review Flows

- After appointment → send satisfaction check + review request
- Missed appointment → send reschedule link
- After first visit → send check-in and suggested next step

Days 26–30 — Review, Refine, and Expand

- Pull 20–50 AI-handled conversations
- Identify what worked and what didn't
- Refine prompts, knowledge base, and rules
- Decide what the next layer of automation should be

By Day 30:

- customers get answers faster
- your phones are a little quieter
- your team is less overwhelmed
- your schedule has fewer gaps and fewer no-shows

You've successfully created a **customer service and scheduling system powered by AI**, enabling your business to deliver fast, reliable support and seamless appointment management. All of this was accomplished without the need to expand your front desk team, allowing you to operate more efficiently and provide a modern experience for your customers.

MOVING FORWARD

Customer Service Is Still Human — AI Just Makes It Possible to Deliver at Scale

AI does not replace empathy.
It does not replace real care.
It does not replace your ability to build relationships.

What it does replace is:

- unanswered messages

- missed calls
- forgotten follow-ups
- manual reminder work
- scheduling chaos

It gives your team:

- space to breathe
- time to focus on high-value problems
- the ability to be truly present with customers who need a human

In an AI-enabled world, the best customer experiences will combine:

- **AI speed and consistency**
- **human warmth and judgment**

You're now equipped to deliver that combination.

Next, we'll extend this same mindset deeper into your business with **AI for Operations & Admin** — where all the invisible, boring, absolutely critical work gets faster, cleaner, and far less painful.

CHAPTER 10: AI for Operations & Admin

How Small Businesses Can Eliminate Bottlenecks, Reduce Labor Costs, and Run 10× More Efficiently With AI

INTRODUCTION

Your Business Doesn't Need More Staff — It Needs Better Systems

Every small business owner eventually hits the same wall:

You're not drowning in work because your business is growing...
You're drowning because your **operations aren't built to scale.**

This is the chapter where that all changes.

Operations and administration used to be the "boring" part of business — the paperwork, the scheduling, the coordination, the endless little tasks no one sees but everyone depends on.

Now?
Operations is the backbone of your competitive advantage.

And AI is the operational rocket fuel that lets small teams perform like companies ten times their size.

In this chapter, you'll learn how AI can:

- automate admin tasks you hate
- eliminate bottlenecks that slow your team
- streamline scheduling, follow-up, and internal coordination
- reduce repetitive labor
- standardize processes so errors disappear
- give you real-time visibility into your business
- free you from the day-to-day chaos that drains your time

This chapter hands you the systems, frameworks, and tools to transform your operations from "barely keeping up" to **predictable, scalable, and radically efficient.**

Let's get to work.

SECTION 1 — Why Operations Is the #1 Area Where AI Creates Massive ROI

Marketing gets the attention.
Sales gets the headlines.
But **AI produces the fastest ROI in operations and admin**, and here's why:

1. Operations are where the most time is wasted

Small businesses lose 10–25 hours per week to tasks like:

- scheduling
- invoicing
- inbox triage
- document creation
- data entry
- follow-up
- reporting
- file management

AI eliminates 50–90% of this.

2. Operations is where errors multiply

Human admin work leads to:

- double bookings
- wrong invoice amounts
- lost messages
- missing information
- missed appointments
- incomplete documentation

AI brings **100% consistency** and catches mistakes before they happen.

3. Operations is where margins disappear

When you're paying employees (or yourself) for repetitive admin work, every hour costs you.

AI cuts:

- admin labor
- back-office waste
- operational drag
- unnecessary software costs
- human bottlenecks

4. Operations is the hardest function to scale manually

You can't grow if you are:

- buried in tasks
- chasing paperwork
- coordinating schedules
- writing the same emails repeatedly

AI gives you unlimited leverage.

SECTION 2 — The Admin Tasks AI Eliminates Immediately

Here's what AI can take off your plate within hours:

1. Email Management & Inbox Triage

AI can:

- categorize emails
- draft replies
- summarize long threads

- identify urgent items
- extract action steps

Tools: ChatGPT, Claude, Superhuman AI

2. Scheduling & Calendar Coordination

AI can:

- book appointments
- reschedule automatically
- send reminders
- sync calendars
- reduce no-shows
- detect double bookings

Tools: Calendly AI, TidyCal, Motion AI

3. Document Creation

AI can draft:

- proposals
- contracts
- SOPs
- training manuals
- intake forms
- job descriptions
- onboarding materials

Tools: ChatGPT, Notion AI, Canva AI, Google Workspace AI

4. Reporting & Analytics

AI can generate:

- weekly summaries
- KPI dashboards
- revenue insights
- forecasting
- customer sentiment analysis

Tools: ChatGPT (data analysis), Google Analytics AI, CRM AI modules

5. Internal Communication

AI can:

- write meeting notes
- extract next steps
- assign tasks
- send team updates
- summarize long Slack threads

Tools: Notion AI, Slack AI, ClickUp AI

6. File & Knowledge Organization

AI can:

- tag documents
- auto-organize shared drives
- convert voice notes into documents

- answer employee questions using your files

Tools: Notion AI, Guru AI, Dropbox AI

SECTION 3 — The 6 Operational AI Systems Every SMB Should Build

These are the systems that transform your business from reactive chaos to proactive efficiency.

SYSTEM 1 — AI Scheduling & Appointment Automation

AI handles:

- booking
- rescheduling
- cancelations
- reminders
- confirmations
- follow-up

Benefits:

- fewer no-shows
- faster responses
- higher booking rates

Tools: Calendly, Tidio, GoHighLevel, Air.ai

SYSTEM 2 — AI Customer Intake & Admin Automation

AI collects:

- customer info

- job details
- service needs
- preferences
- billing details

It then routes data automatically into:

- your CRM
- your inbox
- your reports
- your scheduling system

Tools: Jotform AI, Typeform AI, Zapier

SYSTEM 3 — AI Workflow Automation Engine

When X happens → automatically trigger Y.

Example workflows:

- Lead fills out form → send SMS → notify team → add to CRM
- Customer books meeting → send prep email → add to calendar
- Invoice paid → send receipt → update QuickBooks
- Call ends → AI generates notes → sync to CRM

Tools: Zapier, Make

SYSTEM 4 — AI Document Engine

AI creates, maintains, and updates:

- SOPs
- training guides
- operations manuals
- onboarding docs
- templates

This system standardizes and professionalizes your business.

Tools: ChatGPT, Claude, Notion AI

SYSTEM 5 — AI Reporting & Visibility Dashboard

AI analyzes:

- revenue trends
- customer behavior
- employee performance
- marketing data
- operational bottlenecks
- forecasting

You ask it simple questions like:

"Where did we lose the most time last week?"
"What patterns do you see in this month's revenue?"

Tools: ChatGPT, CRM analytics, Google Analytics AI

SYSTEM 6 — AI Internal Communication Engine

AI turns your team communication into clear action:

- summaries

- assignments
- checklists
- reminders
- meeting notes

Your team stays aligned with zero effort.

Tools: Notion AI, Slack AI, ClickUp AI

SECTION 4 — Industry-Specific Operational AI Workflows

Here's how AI transforms operations across sectors:

Medical Offices & Med Spas

AI handles:

- patient intake
- appointment reminders
- follow-up instructions
- documentation summaries
- consent forms
- aftercare messaging

Home Services (HVAC, Plumbing, Electrical, Landscaping)

AI handles:

- job scheduling
- estimate creation
- route optimization
- work order generation

- follow-up messages
- maintenance reminders

Restaurants

AI handles:

- reservation questions
- menu updates
- staffing schedules
- inventory forecasting
- event coordination

Retail & E-Commerce

AI handles:

- inventory management
- product descriptions
- stock alerts
- vendor communication
- fulfillment coordination

Professional Services (Law, Accounting, Consulting)

AI handles:

- document creation
- client intake
- invoice reminders
- appointment scheduling

- case/project summaries

SECTION 5 — The AI Operations Scorecard

Evaluate your current operations using this simple scorecard.

Rate each category 1–5:

- Scheduling & calendar management
- Inbox & communication
- Admin workload
- Documentation
- Reporting & analytics
- Customer intake
- Task delegation
- Internal communication
- File management
- Data accuracy

If your score is **under 35**, AI can produce massive immediate gains.
If under **25**, its mission critical.

SECTION 6 — Your 30-Day AI Operations Blueprint

Week 1 — Identify Your Top 3 Operational Bottlenecks

Examples:

- scheduling
- slow admin work

- documentation backlog
- missed customer information

Week 2 — Choose Your First Tool

Pick ONE:

- Calendly AI
- ChatGPT
- Zapier
- Notion AI

Week 3 — Build the First Automation

Examples:

- auto-scheduling
- automated follow-up
- automated documentation
- weekly reporting summaries

Week 4 — Measure, Refine, Scale

Track:

- hours saved
- errors reduced
- customer satisfaction
- workload shifts
- financial impact

Then expand one workflow at a time.

SECTION 7 — The Biggest Mistakes Small Businesses Make with AI Operations

X Automating too much too soon
X Using AI without documenting the new process
X Not training one "AI Champion"
X Relying solely on chatbots
X Ignoring data cleanup
X Letting tools overlap or duplicate functions
X Expecting perfection immediately

Fix the process → then automate the process → then scale the system.

MOVING FORWARD

AI Turns Ordinary Operations into a Competitive Superpower

When AI enters your operations:

- work speeds up
- errors disappear
- your team gets smarter
- customers get faster service
- your stress drops
- your margins increase
- your capacity grows
- your business becomes scalable

This is where real transformation happens.

Small businesses who implement AI in operations dominate their markets not because they're bigger...

...but because they're **faster, leaner, more consistent, more accurate, and infinitely more efficient.**

Your operations are the engine of your business.

Now — with AI — it's time to supercharge that engine.

CHAPTER 11: AI for HR & Hiring

How Small Businesses Can Recruit Faster, Hire Better, and Build Stronger Teams in the Age of AI

INTRODUCTION

Hiring Isn't Broken — Your Old Hiring System Is

Ask any small business owner what their #1 frustration is, and you'll hear the same answer:

"I can't find good people."

But the truth is more complicated.

Most small businesses aren't struggling because great employees no longer exist.

They're struggling because:

- their hiring process is outdated
- their job descriptions are weak
- their screening process is slow
- their follow-up is inconsistent
- their interviews aren't structured

- their onboarding is messy
- their employer brand is invisible
- their HR admin drains time
- their documentation is a disaster

AI doesn't magically "fix hiring."
It rebuilds hiring from the ground up—into a predictable, efficient, professional system.

This chapter will show you how to:

- attract better candidates
- respond faster than competitors
- filter applicants more accurately
- eliminate paperwork
- create structured interviews
- improve retention
- onboard employees consistently
- turn HR into a strategic advantage

Hiring is no longer a guessing game.

With AI, **small businesses finally gain access to the recruitment power that only big companies had before.**

Let's build the AI-enabled hiring system your business deserves.

SECTION 1 — Why Hiring Is the Perfect Use Case for AI

Hiring is the most broken part of most small businesses and the part where AI creates instant ROI.

Reason 1 — Hiring Is Mostly Repetitive Admin Work

Posting jobs.
Screening applicants.
Sorting resumes.
Sending follow-ups.
Scheduling interviews.
Creating onboarding docs.
Repeating the same questions over and over.

AI handles 60–80% of this automatically.

Reason 2 — Hiring Requires Fast Response Times

The best candidates disappear in hours, not weeks.

AI lets you:

- respond instantly
- send screening questions
- schedule interviews
- keep candidates engaged

Speed wins talent.

Reason 3 — Hiring Requires Better Messaging

Most SMB job descriptions are:

- vague

- outdated
- uncompetitive
- unclear
- boring

AI helps you write compelling, modern, attention-grabbing job ads that attract high-quality applicants.

Reason 4 — Hiring Requires Consistency

Most SMBs hire reactively:

- different questions every time
- no scoring system
- no structured interviews
- inconsistent onboarding

AI brings structure, clarity, and predictability.

Reason 5 — Hiring Is Expensive When Done Wrong

AI reduces:

- turnover
- mis-hires
- training waste
- onboarding mistakes
- time spent on admin
- compliance errors

This is where small businesses gain enormous savings.

SECTION 2 — The AI Hiring Stack: The 5 Tools You Actually Need

Small businesses don't need enterprise HR suites.
They need **five simple tool categories** that transform hiring immediately.

1. AI Job Description & Employer Brand Engine

Tools: ChatGPT, Claude, Jasper
What it produces:

- modern job descriptions
- salary benchmarks
- benefit summaries
- hiring ads
- employer value propositions
- role scorecards
- interview profiles

2. AI Resume Screening & Candidate Ranking

Tools: BambooHR AI, Rippling AI, Homerun AI
What it does:

- analyzes resumes
- matches candidates to criteria
- identifies red flags
- ranks applicants
- highlights the strongest fits

This alone saves 5–15 hours per hire.

3. AI Communication & Follow-Up Engine

Tools: GoHighLevel, HubSpot AI, Tidio
What it handles:

- application confirmation messages
- screening questions
- interview scheduling
- reminders
- follow-up
- pipeline updates

You never leave candidates waiting again.

4. AI Interview System

Tools: ChatGPT, Notion AI
What it creates:

- structured interview guides
- scoring systems
- scenario-based questions
- skill assessments
- objective evaluation rubrics

This eliminates subjective, inconsistent interviews.

5. AI Onboarding & Documentation System

Tools: Notion AI, Guru AI, Google Workspace AI
What it automates:

- SOP creation
- onboarding guides
- training modules
- 30/60/90 plans
- role expectations
- ongoing performance tracking

This makes new employees into productive team members faster.

SECTION 3 — The AI-Powered Hiring Pipeline (The SMB Blueprint)

This is the hiring system I implement for clients.
It cuts hiring time by 50–70% and dramatically improves candidate quality.

STEP 1 — Define the Role With AI

AI helps you clarify:

- role responsibilities
- required skills
- key outcomes
- performance metrics
- cultural expectations
- compensation range

Prompt:

“Create a role scorecard for a [role], including responsibilities, competencies, KPIs, and success criteria.”

STEP 2 — Write a High-Conversion Job Description

AI produces:

- a compelling title
- a modern intro paragraph
- benefits and culture highlights
- clear responsibilities
- specific qualifications
- salary transparency
- EEOC-friendly language

Your job description now stands out in any job market.

STEP 3 — Create AI-Generated Screening Questions

AI produces a short list of:

- 3–5 knockout questions
- 3–5 skill-based questions
- optional video-response prompts

Candidates who aren’t serious eliminate themselves.

STEP 4 — AI Filters & Ranks Applicants

AI screens resumes for:

- relevant experience
- red flags
- skills alignment
- work history patterns
- culture fit indicators

It then generates a shortlist for you.

STEP 5 — AI Schedules Interviews

AI handles:

- calendar coordination
- sending links
- reminders
- rescheduling

Which means you never lose great candidates to slow scheduling.

STEP 6 — AI Creates Structured Interview Guides

AI produces:

- 10 role-specific questions
- 10 behavioral questions
- scoring rubrics
- culture fit questions
- situational prompts
- evaluation templates

Your hiring becomes objective, not emotional.

STEP 7 — AI Summarizes Interviews

Record your interview notes or transcript.

AI:

- summarizes key points
- highlights strengths
- identifies concerns
- compares candidates
- recommends next steps

This dramatically improves decision quality.

STEP 8 — AI Creates Your Offer Letter

AI drafts:

- compensation package
- benefits summary
- onboarding steps
- role expectations

Professional. Clear. Fast.

STEP 9 — AI Automates Onboarding

AI generates:

- training plan
- checklists
- first-week schedule

- 30/60/90 day plan
- SOPs for the role
- company handbook summaries

Your new hire becomes productive quickly.

SECTION 4 — Industry-Specific AI HR Workflows

Home Services

AI creates:

- technician onboarding manuals
- safety SOPs
- customer communication scripts
- daily job checklists

Medical Offices & Med Spas

AI handles:

- compliance documentation
- patient communication scripts
- staff training modules
- care protocols

Restaurants

AI generates:

- FOH/BOH training
- scheduling
- hiring ads

- onboarding guides
- safety checklists

Retail & E-Commerce

AI manages:

- product training
- customer service scripts
- seasonal staffing guides
- interview scorecards

Agencies & Professional Services

AI handles:

- job descriptions
- interview flows
- onboarding SOPs
- performance metrics

SECTION 5 — The AI HR & Hiring Scorecard

Rate your current hiring system 1–5:

- job descriptions
- candidate flow
- response time
- screening efficiency
- interview structure
- documentation

- onboarding consistency
- employee retention

If your score is:

Under 30 → your hiring system is harming your growth
Under 20 → your hiring system is actively damaging your business

AI fixes this.

SECTION 6 — Your 30-Day AI Hiring Plan

Week 1 — Build Your Role Scorecard

Use AI to define the role clearly.

Week 2 — Create Job Ads & Screening System

Post AI-optimized ads.
Build knockout questions.

Week 3 — Automate Scheduling & Interview Processes

Set up:

- automated follow-up
- automated scheduling
- structured interviews

Week 4 — Automate Onboarding

Create:

- SOPs

- training modules
- new hire checklist
- 30/60/90 plans

Your hiring is now professional, fast, and consistent.

SECTION 7 — Biggest Mistakes SMBs Make with AI in HR

X Using AI to replace judgment
X Automating before you clarify the role
X Copy-pasting generic job descriptions
X Not training AI on company culture
X Overlooking compliance and safety
X Letting AI screen candidates alone
X Neglecting human relationship-building

AI is an amplifier.
It amplifies clarity if you have it — and chaos if you don't.

MOVING FORWARD

AI Doesn't Replace Your People — It Helps You Hire the Right Ones

When done correctly, AI helps small businesses:

- attract better talent
- hire faster
- reduce turnover
- improve culture

- create clarity
- eliminate paperwork
- increase retention
- build strong teams

Hiring shouldn't feel like gambling anymore.

With AI, your small business finally gains the structure, sophistication, and speed of a professional HR department — without the cost of one.

It's time to stop guessing.
It's time to stop reacting.
It's time to hire with confidence — and build the team that will carry your business into the future.

CHAPTER 12: AI for Finance & Accounting

How Small Businesses Can Finally Get Control of Their Numbers, Cash Flow, Pricing, and Profitability Using AI

VALUE PROPOSITION

AI Won't Just Save You Time. It Will Save Your Business from Financial Blind Spots

There is one universal truth about small business finances:

Businesses don't fail because of a lack of passion. They fail because of a lack of financial visibility.

Most owners have no idea what's happening in their numbers until something breaks:

- the bank account sinks
- payroll gets tight
- a tax bill hits
- invoices are past due
- cash flow dries up

- expenses creep up silently
- profit margins shrink

This isn't incompetence.
It's reality.

Most small business owners didn't go into business to manage spreadsheets, reconcile accounts, or analyze financial patterns.

They went into business to *do the work they love.*

But AI changes everything.

AI transforms financial management from a confusing, stressful, reactive chore into a:

- proactive
- automated
- accurate
- predictive
- always-on
- data-driven

financial intelligence system.

This chapter will show you how to:

- automate bookkeeping
- eliminate invoicing delays
- prevent cash flow emergencies
- forecast revenue

- reduce expenses
- spot pricing opportunities
- detect fraud or errors
- simplify taxes
- improve financial decision-making

This is financial clarity at a level SMBs have never had access to before.

Let's build the AI-powered financial engine your business has been missing.

SECTION 1 — Why AI Is a Finance Superpower for SMBs

Reason 1 — AI removes human error from bookkeeping

Most bookkeeping mistakes come from:

- Manual Data Entry
- lost receipts
- miscategorized transactions
- inconsistent tracking
- skipped reconciliations

AI eliminates these error sources instantly.

Reason 2 — AI automates repetitive financial tasks

AI can handle:

- categorizing transactions
- matching invoices

- detecting duplicates
- calculating taxes
- generating P&Ls
- monitoring cash flow
- preparing monthly summaries

What used to take 5–10 hours per week now takes minutes.

Reason 3 — AI gives you real-time visibility

You no longer wait until the end of the month (or quarter) to "see the numbers."

AI gives you:

- daily snapshots
- weekly summaries
- monthly reports
- predictable cash flow projections

You finally run your business with financial intelligence — not guesswork

Reason 4 — AI is cheaper than a part-time bookkeeper

AI tools cost:

$20–$60/month

Meanwhile, bookkeepers cost:

$300–$2,000/month

AI doesn't replace bookkeepers — but it reduces manual workload by 70% so they can focus on strategy, not data cleaning.

Reason 5 — AI helps you make smarter, faster decisions

AI can analyze:

- your revenue patterns
- your busiest days
- your slow seasons
- your cost leaks
- your pricing gaps
- which products/services are most profitable
- which employees produce best margins

This is the kind of insight that used to cost tens of thousands in consulting.

SECTION 2 — The AI Finance Stack: The Tools That Actually Matter

You don't need dozens of tools.
You need five categories.

1. AI Bookkeeping Automation

Tools: QuickBooks AI, Wave AI, Xero AI

AI handles:

- transaction categorization

- reconciliation suggestions
- expense organization
- duplicate detection
- receipt extraction
- automated matching
- bank rule creation

This is foundational.

2. AI Invoicing & Billing Automation

Tools: Stripe, Square, FreshBooks AI

AI handles:

- generating invoices
- sending reminders
- tracking payments
- applying late fees
- subscription billing
- predicting missed payments

No more forgetting to bill.
No more waiting 90 days for customers to pay.

3. AI Cash Flow & Forecasting

Tools: QuickBooks AI, Pulse AI, Float AI

AI predicts:

- cash shortages

- slow seasons
- hiring feasibility
- investment timing
- revenue patterns

This lets owners plan *before* problems appear.

4. AI Reporting & Insights

Tools: ChatGPT (upload spreadsheets), PowerBI AI, Google Sheets AI

AI generates:

- weekly summaries
- revenue trends
- cost reduction opportunities
- margin reports
- KPI dashboards

This turns you into a financially proactive CEO.

5. AI Fraud & Error Detection

Tools: AI anomaly detectors inside accounting software

AI flags:

- unusual spending
- duplicate transactions
- suspicious vendor activity
- incorrect categorizations

- payroll inconsistencies

AI protects your business from financial risk better than manual oversight ever could.

SECTION 3 — The 10 Finance Tasks You Should Automate First

These are the easiest, highest ROI finance automations for small businesses.

1. Automatic Transaction Categorization

AI learns your patterns and categorizes with 95%+ accuracy.

2. Automatic Receipt Processing

Snap a photo → AI extracts the data → categorizes → stores → syncs to your books.

3. Automatic Invoice Creation

When a job finishes → AI generates an invoice → emails it instantly.

4. Automatic Payment Reminders

Stop chasing customers manually.
AI does it professionally and consistently.

5. Auto-Generated Monthly Financial Reports

AI creates:

- P&L

- balance sheet
- cash flow
- KPI review

Fully automated.

6. AI Cash Flow Alerts

"Warning: Your cash balance may drop below $5,000 on March 12."

Game-changing.

7. AI Payroll Prep

AI collects hours, organizes data, flags errors.

8. AI Tax Preparation Assistance

AI organizes:

- deductions
- categorized expenses
- mileage logs
- receipts

Your CPA will love you.

9. AI Vendor & Expense Optimization

AI suggests:

- which subscriptions to cancel
- cheaper alternatives

- duplicate services
- spending trends

10. AI Pricing Optimization

AI analyzes:

- competitor pricing
- customer patterns
- demand curves
- product/service profitability

It recommends smarter price adjustments — backed by data.

SECTION 4 — Industry-Specific AI Finance Workflows

Home Services

AI handles:

- job costing
- estimating
- invoicing
- follow-up reminders
- seasonal forecasting

Med Spas & Medical Offices

AI handles:

- insurance summaries
- billing coding guidance
- payment reminders

- patient invoicing
- cost-per-treatment analysis

Restaurants

AI handles:

- ingredient cost forecasting
- labor cost optimization
- menu profitability analysis
- daily revenue summaries

Retail & E-Commerce

AI handles:

- inventory valuation
- stock forecasting
- margin optimization
- vendor analysis

Professional Services

AI handles:

- project-based billing
- retainer forecasting
- client profitability ranking

SECTION 5 — The AI-Driven Finance Dashboard (Your New Superpower)

Every SMB should ask AI every week:

- “How did we perform financially this week?”
- “What patterns should I be concerned about?”
- “Where am I overspending?”
- “Which products/services are most profitable?”
- “What are your predictions for next month’s cash flow?”
- “What financial risks do you detect?”

This replaces confusion with clarity.

SECTION 6 — The 30-Day AI Finance Transformation Plan

Week 1 — Automate Bookkeeping

Set up:

- QuickBooks AI (or equivalent)
- recurring rules
- automatic categorization
- receipt uploads

Week 2 — Automate Billing & Invoices

Set up:

- invoice templates
- reminders
- auto-charges
- late fees
- follow-up sequences

Week 3 — Build Your Financial Dashboard

Have AI generate:

- weekly summary
- monthly KPI report
- cash flow forecast

Week 4 — Integrate Finance Into Operations

Connect finance data to:

- scheduling
- CRM
- payroll
- expenses
- marketing attribution

Now your entire business talks to your numbers.

SECTION 7 — The Biggest Mistakes to Avoid

X Not reviewing AI-generated reports
X Automating before cleaning up existing data
X Relying solely on AI without CPA oversight
X Using multiple finance tools that overlap
X Underpricing services based on emotion instead of AI insights
X Ignoring cash flow predictions

AI doesn't eliminate your responsibility.
It eliminates your *busywork* so you can focus on real decisions.

MOVING FORWARD

AI Turns Your Numbers Into a Strategic Advantage rather than a Stress Point

For the first time in history, small businesses can manage their finances with:

- clarity
- confidence
- real-time intelligence
- predictive insight
- automation
- professional accuracy
- strategic decision-making

You no longer have to guess.
You no longer have to drown in paperwork.
You no longer have to run your business "from the gut."

AI gives you the power of a CFO — without the cost of one.

Your financial future becomes visible.
Controllable.
Predictable.
Optimizable.
Strategic.

This is the chapter where business owners go from overwhelmed...

to informed.
From reactive...
to proactive.
From stressed...
to empowered.

Your numbers finally work *for* you.

CHAPTER 13: The AI-Enabled Workforce

How AI Will Transform Jobs, Teams, and the Way Small Businesses Work — and How to Prepare Your People for What Comes Next

VALUE PROPOSITION

Your Team Isn't Afraid of AI — They're Afraid of Being Left Behind

There is a quiet fear running through small businesses right now.

Employees see AI everywhere — on the news, in their tools, in customer expectations, in conversations with friends. They're not afraid of AI taking their jobs...

They're afraid of getting replaced by **someone else** who *knows how to use AI better.*

This is the truth business owners must understand:

AI doesn't eliminate jobs — it eliminates tasks.
But employees who don't learn AI will be replaced by employees who do.

This chapter shows you how to build an AI-enabled workforce where:

- employees become *more valuable*, not less
- roles evolve instead of disappear
- training becomes continuous
- productivity skyrockets
- stress decreases
- burnout fades
- people feel supported, not threatened
- hiring becomes easier
- retention improves

AI-enabled companies are not just more productive — they are *better places to work.*

This is the future of the small-business workforce.

Let's build it.

SECTION 1 — The Truth About AI and Jobs: What SMB Owners Must Know

There are three forces happening simultaneously:

1. AI removes repetitive, low-value tasks

These include:

- writing repetitive emails
- data entry

- scheduling
- document drafting
- summarizing information
- inbox triage
- transcribing
- basic analysis
- repetitive customer communication

These tasks don't need humans anymore — and humans never liked them.

2. AI amplifies the value of skilled employees

When an employee uses AI effectively, they become:

- faster
- more accurate
- more strategic
- more creative
- more organized
- more consistent

Your best employees become *10× employees*.

3. AI shifts job descriptions without eliminating jobs

Roles evolve.

Examples:

- “Receptionist” → “Customer Experience Coordinator with AI Support”
- “Marketing Assistant” → “Content Strategist with AI Tools”
- “Project Manager” → “AI-Enhanced Operations Leader”
- “Estimator” → “Estimator With AI Workflows”

AI doesn’t wipe out the workforce — it **upskills the workforce.**

SECTION 2 — The Three Types of Employees in the AI Era

Every small business now has these three employee categories:

1. The Multipliers

These employees embrace AI.

They:

- learn fast
- experiment
- document workflows
- think proactively
- help others
- boost productivity
- become leaders

These are your future managers and team leads.

2. The Maintainers

These employees use AI when told, but don't innovate.

They maintain consistency, but not improvement.

They need:

- training
- support
- repetition
- encouragement

With coaching, they can become multipliers.

3. The Resisters

These employees refuse to use AI.

They:

- fear it
- resent it
- avoid it
- believe it's "not part of their job"
- slow down the team
- push work onto others
- sabotage efficiency

These employees either adapt...
or they eventually get replaced by someone who will.

SECTION 3 — The AI Workforce Ladder™

A simple framework for developing AI-enabled employees

There are four levels:

Level 1 — Awareness

Employee understands:

- what AI is
- how the company uses it
- the benefits
- the expectations

Goal: remove fear, build curiosity.

Level 2 — Adoption

Employee uses AI tools for:

- writing
- summarizing
- communication
- documentation
- simple workflows

Goal: reduce repetitive tasks by 30–50%.

Level 3 — Integration

Employee integrates AI into:

- daily workflows
- customer communication

- planning
- problem-solving
- follow-up
- reporting

Goal: reduce repetitive tasks by 50–80%.

Level 4 — Leadership

Employee leads:

- SOP creation
- training
- AI tool implementation
- workflow design
- team coaching

Goal: AI becomes part of your culture.

SECTION 4 — How to Train Your Team to Use AI (Step-by-Step)

This is the system I use with consulting clients.

STEP 1 — Start With a 60-Minute AI Workshop

Cover:

- what AI is
- what AI isn't
- how AI helps them
- how AI affects workload

- how AI reduces stress
- how AI affects job security (in reality)

Outcome: clarity and buy-in.

STEP 2 — Create 5–10 Approved AI Use Cases

Examples:

- writing customer emails
- summarizing meeting notes
- drafting social posts
- creating SOPs
- generating documentation
- answering FAQ
- preparing job descriptions
- formatting spreadsheets

Employees must know exactly what's allowed and what isn't.

STEP 3 — Create Your "AI Prompt Library"

This is one of the most powerful tools in this chapter.

Include prompts for:

- customer replies
- scheduling
- SOP creation
- social content
- reporting

- employee communication
- product descriptions
- task planning
- training guides

This instantly increases the team's output.

STEP 4 — Assign an "AI Champion"

This is a key role.

Responsibilities:

- testing new tools
- training team members
- documenting workflows
- reviewing AI output
- maintaining prompts
- guiding implementation

This person becomes your internal AI leader.

STEP 5 — Evaluate Employees Based on AI Adoption

Add AI competency to:

- performance reviews
- job descriptions
- promotions
- raises
- role expectations

Make AI part of the culture.

SECTION 5 — AI and Culture: How to Avoid Fear, Resistance & Misunderstanding

AI can either strengthen culture...
or break it.

Here's how to do it right.

1. Communicate Early and Often

Employees fear what they don't understand.

Explain:

- what AI will automate
- what it won't
- why you're using it
- how it benefits them
- what's changing
- what's staying the same

2. Emphasize That AI Helps Them — Not Replaces Them

Sample message:

"AI won't take your job.
It will take the tasks that make your job harder."

3. Build a Collaborative, Not Competitive, AI Culture

Avoid making people feel "left behind."

Encourage:

- peer learning
- shared prompts
- team workshops
- public wins

4. Reward Initiative

Recognize employees who:

- streamline workflows
- build new systems
- improve customer experience using AI

This builds momentum.

5. Train Managers First

Managers must model AI usage.

If managers resist, the team will resist.

SECTION 6 — The Future of Work for SMBs (What's Coming Next)

Here is what will happen over the next 3–5 years:

1. Every employee will have an AI assistant

It will:

- plan their day
- draft their tasks

- manage follow-up
- monitor KPIs
- remind them of deadlines

2. AI Agents will take over full workflows

Scheduling
Follow-up
Documentation
Reporting
Inventory
HR tasks
Finance tasks

3. Job descriptions will evolve dramatically

Roles will require:

- communication
- problem-solving
- prompting
- workflow thinking
- soft skills
- adaptability

4. Employees will expect AI training (and prefer AI-enabled workplaces)

If you don't offer AI support:

- retention drops
- recruiting becomes harder

- your culture weakens

5. AI will become the new "literacy"

Just like computers in the 2000s.
Just like smartphones in the 2010s.

AI literacy will be expected in every role.

SECTION 7 — The AI Workforce Scorecard

Rate each area from 1–5:

- employee AI usage
- prompt literacy
- workflow integration
- training consistency
- cultural acceptance
- manager adoption
- documentation
- speed of execution

Score meaning:

30–40: AI-enabled workforce
20–29: partially enabled — big opportunity
10–19: vulnerable to competition
Under 10: at high risk of falling behind

SECTION 8 — Your 30-Day AI Workforce Transformation Plan

Week 1 — Communicate the Vision

Run a kickoff meeting.

Week 2 — Train the Team

Run a 60-minute workshop.
Share the prompt library.

Week 3 — Assign an AI Champion

This person leads internal adoption.

Week 4 — Redesign 3–5 Job Descriptions

Integrate AI expectations.

Your company now operates like a modern business.

MOVING FORWARD

Your Workforce Won't Be Replaced by AI — But It Will Be Transformed by It

The small businesses that thrive over the next decade will not be the ones who simply "use AI."

It will be the businesses that:

- **train their people**
- **reshape their roles**
- **build collaborative AI cultures**
- **reward initiative**

- **eliminate low-value work**
- **free employees to do higher-value work**

This is how you build a team that is:

- faster
- smarter
- more capable
- more confident
- less stressed
- more loyal
- more innovative

The AI-enabled workforce is not a threat.
It's the greatest upgrade small businesses have had in decades.

And now you know exactly how to build it.

CHAPTER 14: AI Safety, Compliance & Trust

How to Use AI in Your Business Without Risking Data, Reputation, or Customer Confidence

VALUE PROPOSITION

AI Isn't Dangerous — Using AI Carelessly Is

Small business owners are right to be excited about AI.
They're also right to be cautious.

AI is powerful.
AI is fast.
AI is everywhere.

But AI is also **misunderstood** and that misunderstanding can lead to:

- data exposure
- privacy violations
- inconsistent messaging
- inaccurate content
- compliance mistakes

- damaged customer trust

The danger isn't AI itself.
The danger is:

- *using AI with no guidelines*
- *letting employees experiment without rules*
- *uploading sensitive data into public models*
- *assuming AI is always correct*

This chapter gives you **the exact safety framework** used by the largest companies in the world and we've simplified it for small businesses.

You will learn how to:

- protect customer data
- stay compliant
- prevent "AI hallucinations"
- train employees safely
- choose trustworthy tools
- build an AI policy
- control access and permissions
- avoid scams and risky automations
- maintain your brand voice
- build long-term customer trust

AI safety isn't complicated — when you know what you're doing.

Let's build the guardrails that keep your business protected.

SECTION 1 — What AI Safety Actually Means (SMB Version)

Forget the overhyped headlines about robots, super-intelligence, or job extinction.

For small businesses, **AI safety** means five things:

1. Data Safety

Protecting customer information, financial records, personal details, and internal documents.

2. Compliance Safety

Following:

- GDPR
- HIPAA
- SOC2
- PCI
- local privacy laws
- industry regulations

You don't need to be a lawyer. AI just needs guardrails.

3. Brand Safety

Ensuring AI:

- uses the right tone
- doesn't invent facts
- doesn't mislead customers
- doesn't break brand standards

4. Operational Safety

AI should *assist* humans and not make independent decisions on sensitive tasks.

5. Reputation Safety

Customers must trust your systems.

If you implement AI responsibly, trust increases.

SECTION 2 — The 7 Most Common AI Risks (And How to Prevent All of Them)

These are the real risks small businesses face.

And every one of them is easily preventable.

RISK 1 — Uploading Sensitive Information into Public AI Models

Never paste:

- customer addresses
- credit cards
- medical records
- legal documents
- payroll info
- employee data
- confidential IP

Fix:
Use business-tier AI tools with data privacy protections.

RISK 2 — AI "Hallucinations" (Incorrect Answers)

AI sometimes generates:

- wrong facts
- made-up citations
- incorrect numbers
- bad advice
- confident but false explanations

Fix:
Always verify content before publishing or sending.

RISK 3 — Inconsistent Brand Voice

AI can produce:

- robotic replies
- overly formal text
- unprofessional tone

Fix:
Create a **Brand Voice File** and feed it to the model.

RISK 4 — Over-Automation

Automating:

- refunds
- pricing decisions
- payroll
- policy exceptions

- sensitive customer situations

...without human review is dangerous.

Fix:
Human-in-the-loop for all sensitive tasks.

RISK 5 — Tool Overload

Using too many AI tools increases:

- cost
- confusion
- data exposure
- operational complexity

Fix:
Use a small, vetted AI tech stack.

RISK 6 — Employee Misuse

Employees experiment with AI unsafely when:

- they're not trained
- no rules exist
- expectations are unclear

Fix:
Implement a written **AI Usage Policy** (template included in this chapter).

RISK 7 — Scam Tools & Fake AI Startups

There are thousands of:

- unregulated tools
- copycat apps
- data-harvesting "AI bots"
- plugins with hidden permissions

Fix:
Stick to trusted platforms (listed later in this chapter).

SECTION 3 — The AI Safety Shield

A Simple Framework for Protecting Your Business

This 4-part framework prevents 99% of AI risks.

1. Protect Your Data

Choose tools with:

- no data retention
- enterprise-level encryption
- SOC2 compliance
- HIPAA support (medical)
- transparent data policies

Approved categories:

- ChatGPT Team / Enterprise
- Claude for Work
- Microsoft 365 Copilot
- Google Workspace Gemini
- Notion AI Business

- Tidio AI (secure SaaS)
- HubSpot AI / GoHighLevel AI

2. Protect Your Workflow

Use human review for:

- pricing
- legal content
- medical guidance
- invoices
- contracts
- customer disputes
- payroll

AI drafts. Humans approve.

3. Protect Your Brand

Create guardrails:

- "Our tone can never..."
- "We do not make guarantees about..."
- "Avoid these phrases..."
- "Always include disclaimers..."

Train your AI with:

- Brand Voice File
- Standard messaging
- Examples of your style

- Competitor differentiation

4. Protect Your Customers

Make AI:

- fast
- accurate
- honest
- transparent

Never let AI:

- argue
- guess
- pretend
- make promises
- hide limitations

This protects customer trust.

SECTION 4 — The AI Safety Policy Every SMB Needs

Below is the **exact policy** I implement for clients — easy to copy into your handbook.

SECTION A — Allowed AI Use Cases

Employees *can* use AI for:

- writing drafts
- summarizing content
- research

- SOP creation
- documentation
- customer FAQ
- marketing content
- internal communication
- task planning

SECTION B — Prohibited AI Use Cases

Employees *cannot* use AI for:

- uploading sensitive data
- sending unreviewed messages
- making financial decisions
- customer refunds
- legal guidance
- medical advice
- policy exceptions

SECTION C — Privacy Requirements

Employees must:

- remove customer identifiers
- redact personal info
- avoid external upload of confidential docs
- use company-approved AI tools only

SECTION D — Brand Requirements

AI content must:

- match brand tone
- be reviewed before use
- follow messaging guidelines
- reflect company values

SECTION E — Human Oversight

Every AI output must be:

- reviewed
- edited
- approved
- checked for accuracy

SECTION F — Tool Approval Process

No employee may add an AI tool without:

- manager approval
- security review
- workflow integration check
- official documentation

SECTION 5 — AI Compliance: What SMBs Need to Know

AI compliance is straightforward when broken down into three categories:

CATEGORY 1 — Local Privacy Laws

Depending on your state/country:

- GDPR (Europe)

- CCPA / CPRA (California)
- PIPEDA (Canada)
- HIPAA (medical)

Most relate to data handling — AI tools simply need to respect these rules.

CATEGORY 2 — Industry Regulations

Examples:

- healthcare documentation
- financial reporting
- legal confidentiality
- insurance compliance
- educational privacy

AI can support compliance — but not replace professional oversight.

CATEGORY 3 — Vendor Compliance

Use tools that are:

- SOC2 certified
- GDPR compliant
- encrypted
- transparent

If a tool does not clearly state its compliance level, do not use it.

SECTION 6 — How to Build Customer Trust in an AI-Driven Business

AI can increase trust *if your customers know how you use it.*

Here's how to build transparency:

1. Tell customers when they're talking to AI

...but emphasize human oversight.

2. Make your AI support fast, accurate, and respectful.

3. Give customers the option to switch to a human anytime.

4. Never use AI to deny, argue, or negotiate.

5. Use AI for consistency and use humans for empathy.

This is how you strengthen your brand through technology.

SECTION 7 — The AI Safety Red Flags Checklist

Before you use any AI tool, ask:

- Does it explain how my data is used?
- Can I delete my data easily?
- Does it store prompts?
- Does it train on my private info?
- Does it offer a business/compliance plan?
- Is the company legitimate?
- Is the interface secure (https)?
- Does it limit employee access?

If any answer feels "off," skip the tool.

SECTION 8 — The 30-Day AI Safety Implementation Plan

Week 1 — Audit Your Current Tools

Identify risky apps.

Week 2 — Create Your AI Safety Policy

Train managers first.

Week 3 — Transition to Secure Tools

Migrate:

- ChatGPT → Team or Enterprise
- Claude → Work
- Google Workspace → Gemini
- Microsoft 365 → Copilot

Week 4 — Train Employees

Hold a 45-minute session covering:

- what AI is
- what's allowed
- what's prohibited
- how to handle customer info

Your business is now protected.

MOVING FORWARD

AI Safety Isn't About Fear — It's About Leadership

AI safety is not complicated.
It's not expensive.
It's not technical.

It's simply:

- clear policies
- smart tool selection
- proper oversight
- respectful communication
- responsible data handling

Small businesses that adopt AI *responsibly* will dominate their industries because trust becomes a differentiator.

Safety is not a limitation.
It's a competitive advantage.

And now you have everything needed to use AI with confidence, clarity, and control.

CHAPTER 15: Choosing the Right AI Tools Without Getting Scammed

A Field Guide for Small Businesses Navigating the Wild West of AI

VALUE PROPOSITION

The AI Tool Boom: More Options, More Confusion, More Risk

AI is the biggest technology land rush since the early internet — and just like the early internet, it's crowded with:

- brilliant tools
- useless tools
- overpriced tools
- half-baked ideas
- cloned products
- shady subscription traps
- fake "AI gurus"
- aggressive upsells
- tools that vanish overnight

Small business owners are especially vulnerable because:

- you don't have time to test everything
- you don't have a CTO who vets software

- you don't have money to burn
- you can't afford downtime or broken systems
- you rely on tools that must *actually* work

This chapter gives you a **battle-tested, scam-proof evaluation system** used by AI consultants, enterprise buyers, and technical reviewers. It has been rewritten for the real world of small business.

By the end of this chapter, you will know:

- what to buy
- what to avoid
- how to evaluate tools in minutes
- how to avoid scam products
- how to avoid overpriced subscriptions
- how to choose tools that will LAST
- how to protect your business from data risks
- how to build a lean, cost-effective AI stack

Let's turn the chaos of the AI tool boom into a structured decision-making advantage.

SECTION 1 — The Harsh Truth: 90% of AI Tools Aren't Worth Your Time

Here's the reality nobody wants to say out loud:

Most AI tools are nothing more than shiny wrappers around ChatGPT.

You'll see tools promising:

- "AI sales copy"
- "AI email writing"
- "AI idea generation"

- "AI business analysis"
-

But when you look under the hood?

They're literally sending your prompts to ChatGPT... and charging you $49–$199/month for the privilege.

You can do the same thing *yourself* for $20.

If a tool is just ChatGPT in a cute outfit, don't buy it.
This chapter teaches you how to spot these instantly.

SECTION 2 — The AI Buyer's Pyramid (What to Adopt First)

Here is the correct order for building an AI tool stack:

LEVEL 1 — Essential (Everyone Needs These)

1. **AI writing/analysis model**
   – ChatGPT or Claude

2. **AI automation tool**
   – Zapier or Make

3. **AI customer response tool**
   – Tidio, Intercom, or Air.ai

LEVEL 2 — High ROI For Most Businesses

4. **AI CRM with automation**
   – HubSpot, GoHighLevel, Close

5. **AI marketing/design**
   – Canva AI + Jasper

LEVEL 3 — Optional, Based on Industry

6. **AI scheduling** (Calendly)
7. **AI bookkeeping** (QuickBooks AI)
8. **AI HR/onboarding** (Rippling or BambooHR)
9. **Industry tools** (Shopify Magic, Toast AI, DrChrono)

LEVEL 4 — Avoid Unless You're a Tech Pro

10. **Brand-new tools with no funding, no track record, and tiny teams**
11. **Tools that duplicate features you already have**
12. **"AI everything" bundles that promise the world for $49/mo**

This pyramid protects your budget **and** accelerates ROI.

SECTION 3 — The 10-Minute Tool Evaluation System (TTES)

A simple, bulletproof process to avoid 99% of bad tools.

Set a timer.
Evaluate any AI tool using these ten criteria:

CRITERIA 1 — Does this tool solve a *real* workflow or just sell hype?

Good: "Automate your lead follow-up."
Bad: "10x your pipeline with quantum AI automation engines!"

If it sounds like a motivational poster, walk away.

CRITERIA 2 — Is this something ChatGPT can do better for cheaper?

If yes → don't buy it.
Period.

CRITERIA 3 — Does the tool replace or integrate with your existing systems?

Good: "Integrates with CRM."
Bad: "Export CSVs manually every time."

If it increases workload, it's worthless.

CRITERIA 4 — Does the company have real funding or a legitimate team?

Red flags:

- no LinkedIn presence
- no team page
- anonymous founders
- no investors

These tools disappear overnight.

CRITERIA 5 — Is customer support real or AI-only?

If you can't reach a human → don't rely on the tool for mission-critical work.

CRITERIA 6 — Does the tool have a clear pricing structure?

Scam signals:

- "limited lifetime offer for $497"

- complicated usage-based fee traps
- no refund policy

Good tools are simple and transparent.

CRITERIA 7 — Does the company explain how your data is handled?

If they can't answer this question:

"Where does my data go and how is it stored?"

Leave.

CRITERIA 8 — Is there a free trial?

If not → the company knows you wouldn't buy it if you tried it.

CRITERIA 9 — Does the tool have real user reviews outside their website?

Check:

- G2
- Capterra
- Reddit
- YouTube reviewers

Fake tools inflate reviews on their own site.

CRITERIA 10 — Does the tool produce measurable ROI in 14 days?

If it can't pass this test:

Save time
Save money
Increase leads
Reduce labor
Improve accuracy

...don't buy it.

SECTION 4 — The Most Common AI Scams (And How to Spot Them Instantly)

SCAM 1 — The "Lifetime Deal" Trap

Any AI tool offering a lifetime license is hiding something:

- they have no funding
- they expect to shut down
- they are reselling someone else's API
- they won't be around next year

Avoid.

SCAM 2 — Tools That Sell "Proprietary Models" They Don't Actually Have

If you see phrases like:

- "Our breakthrough algorithm..."
- "Our proprietary large language system..."
- "Our exclusive AI engine..."

...but they're a tiny startup with 4 employees?

They're lying.
They're using OpenAI's API.

SCAM 3 — "AI Agencies" That Are Really Fiverr Resellers

Signs:

- no portfolio no registered business
- no team page
- vague deliverables
- heavy pressure to buy

Many will just paste your requests into ChatGPT.

SCAM 4 — Tools With No Company Behind Them

If you can't find:

- founders
- office location
- team members
- real contact info

Run.

SCAM 5 — Fake "AI Gurus" Selling Courses

Red flags:

- screenshots of imaginary revenue
- "$10K/month passive AI agent system!"
- "Just resell AI tools!"
- countdown timers that restart when refreshed

Ignore them.
They are not helping businesses — they are milking them.

SECTION 5 — The 7 Rules of Buying AI Tools Smartly

RULE 1 — Start With Pain, Not Hype

If there's no pain, there's no need.

RULE 2 — Buy One Tool Per Category

More tools = more chaos.

RULE 3 — Avoid Tools Younger Than 12 Months

The failure rate is enormous.

RULE 4 — Don't Buy Tools That Duplicate CRM Features

Your CRM probably already does it.

RULE 5 — Avoid Tools With No Integrations

A tool that doesn't connect is a tool you shouldn't use.

RULE 6 — Do a 14-Day ROI Test

If you can't measure impact quickly, the tool is wrong.

RULE 7 — Never Buy Annual Plans First

Tools update monthly — lock-in is dangerous.

SECTION 6 — The AI Tool Decision Chart (What to Buy, When, and Why)

If you want more leads → buy customer response automation

(Tidio, Intercom, Air.ai)

If you want more consistent marketing → buy Canva + Jasper

If you want faster operations → buy Zapier or Make

If you want better sales → buy HubSpot or GoHighLevel

If you want better decisions → buy ChatGPT/Claude premium

If your industry has specialized AI → use that after the fundamentals

SECTION 7 — The 30-Day AI Tool Implementation Plan

WEEK 1 — Identify 2 workflows
WEEK 2 — Pick 2 tools (max)
WEEK 3 — Train 1 employee
WEEK 4 — Automate 1 workflow per tool

At the end of 30 days, you have a lean, powerful, safe AI stack — with no waste.

MOVING FORWARD

Your AI Tools Will Either Supercharge Your Business — or Drain It

The goal of AI is not to collect tools.
The goal is:

- speed
- clarity
- accuracy
- consistency
- automation
- growth

When you buy slowly, intelligently, and strategically...
AI becomes the most valuable employee you've ever had.

When you buy emotionally, impulsively, or without evaluation...
AI becomes an expensive distraction.

You now have the system to choose correctly — every time.

CHAPTER 16: Your 12-Month AI Roadmap

How to Build a Modern, Efficient, AI-Powered Business Step-By-Step — Without Overwhelm, Waste, or Confusion

INTRODUCTION

The Future Isn't Built in a Weekend — It's Built in a Roadmap

Most business owners make two critical mistakes when they try to adopt AI:

Mistake #1: They try to do too much in the first 30 days.
Mistake #2: They have no plan for months 2 through 12.

The result?

- Burnout
- Confusion
- abandoned tools
- wasted time
- wasted money
- a belief that "AI didn't work"

But the problem isn't AI.
The problem is the approach.

Winning with AI is not about intensity.
It's about sequence.

It's about momentum.
It's about building the right system in the right order.

This chapter gives you the exact **12-month roadmap** I use with clients across:

- home services
- medical practices
- real estate teams
- retail shops
- restaurants
- e-commerce
- professional services
- construction
- hospitality
- coaches and consultants

This is the most practical and accessible yearlong AI transformation plan ever written for small businesses.

Let's build your AI-powered year.

SECTION 1 — The Four Phases of AI Transformation

Every business goes through four stages:

PHASE 1 — Foundation (Months 1–3)

Learn, pilot, stabilize.
Small experiments → big confidence.

PHASE 2 — Expansion (Months 4–6)

Scale what works.
Add automation.
Train employees.

PHASE 3 — Optimization (Months 7–9)

Remove bottlenecks.
Tune operations.
Improve accuracy and consistency.

PHASE 4 — Acceleration (Months 10–12)

Build full-speed AI operations.
Multiply growth.
Stand out in your market.

The following roadmap walks you through all four — step by step, in business-friendly language.

SECTION 2 — Months 1–3: Foundation (The "AI Stability Phase")

This is where you:

- pick your tools
- learn how AI behaves
- build your first wins
- automate your first workflow
- reduce your workload
- create internal confidence

MONTH 1 — Start Small, Win Fast

Your milestones:

1. Choose ONE workflow
2. Choose ONE tool
3. Implement ONE automation

4. Track ROI

5. Train ONE employee

Deliverables:

- ✓ AI writing model (ChatGPT or Claude)
- ✓ AI workflow #1 automated
- ✓ AI prompt library started
- ✓ AI champion identified on your team

MONTH 2 — Build Your Core AI Stack

Add only **one** tool per category.

1. Customer response AI (Tidio, Intercom, Air.ai)

2. AI CRM automation (HubSpot or GoHighLevel)

3. AI marketing support (Canva AI or Jasper)

Deliverables:

- ✓ instant lead response
- ✓ automated follow-up
- ✓ AI-assisted content creation
- ✓ brand voice trained into AI
- ✓ basic Zapier workflows

MONTH 3 — Create Consistency Systems

This month creates business stability.

Your goals:

- build an AI-powered content calendar

- create AI templates for email, SMS, and scripts
- automate review requests
- automate basic scheduling

Deliverables:

- ✓ 30 days marketing content done
- ✓ AI powered customer messaging library
- ✓ lead follow-up automation
- ✓ review request automation
- ✓ scheduling assistance

By the end of Month 3?

You have **momentum.**

You have **confidence.**

You have **proof AI works.**

SECTION 3 — Months 4–6: Expansion (The "AI Scaling Phase")

Now that you have traction, it's time to scale what works.

MONTH 4 — AI for Operations

Your goals:

- eliminate repetitive admin
- reduce employee workload
- improve accuracy
- streamline processes

Focus areas:

- invoices
- scheduling
- project management
- documentation
- inventory

- daily reports

Deliverables:

- ✓ 3–5 operations workflows automated
- ✓ internal SOPs drafted using AI
- ✓ employee performance reporting via AI

MONTH 5 — AI for Marketing & Lead Generation

This month is revenue-focused.

Deliverables:

- ✓ industry-specific campaigns
- ✓ automated lead nurturing
- ✓ a 90-day content engine
- ✓ website chat automation
- ✓ appointment booking automation
- ✓ consistent Google Business posts via AI

Add:

- short-form video scripts
- ad variations
- seasonal campaigns

This month typically increases leads by 20–50%.

MONTH 6 — AI for Sales & Customer Experience

Now you strengthen every customer interaction.

Your objectives:

- eliminate slow replies
- eliminate follow-up gaps
- personalize messaging
- improve closing rates

Deliverables:

- ✓ sales scripts
- ✓ AI-assisted proposals
- ✓ objection-handling sequences
- ✓ SMS/email nurturing
- ✓ CRM-driven lead scoring

This month often results in a **noticeable revenue lift**.

SECTION 4 — Months 7–9: Optimization (The "AI Fine-Tuning Phase")

Now your business shifts from "AI adoption" to "AI mastery."

MONTH 7 — Review, Repair, Replace

Identify:

- what's working
- what's not
- what needs improvement
- what can be eliminated

Deliverables:

- ✓ tool consolidation (remove duplicates)
- ✓ improved automations
- ✓ updated prompt library
- ✓ improved templates
- ✓ better data cleanliness

MONTH 8 — Build Your AI Knowledge Base

Turn your business into a documented machine.

Use AI to generate:

- SOPs
- training manuals
- onboarding guides
- internal FAQs
- customer FAQs
- employee workflows

Deliverables:

- ✓ 20–50 pages of SOPs from AI
- ✓ onboarding automation
- ✓ training workflows
- ✓ documented processes

This increases scalability and reduces owner workload dramatically.

MONTH 9 — AI-Driven Analytics & Optimization

This month transforms guesswork into data.

AI will now:

- summarize marketing performance
- analyze sales trends
- measure response times
- highlight customer patterns
- identify inefficiencies

Deliverables:

- ✓ weekly AI business dashboards
- ✓ monthly AI analytics reviews
- ✓ revenue trend analysis
- ✓ cost analysis
- ✓ forecasting assistant

Owners typically see **new clarity** they've never had before.

SECTION 5 — Months 10–12: Acceleration (The "AI Advantage Phase")

This is where your business becomes *unfairly competitive.*

MONTH 10 — Build AI Agents for Key Workflows

Start adding "mini employees" powered by AI.

Examples:

- AI receptionist
- AI customer service rep
- AI sales assistant
- AI marketing coordinator
- AI HR/onboarding assistant

Deliverables:

✓ 2–3 AI agents active
✓ human review systems in place
✓ measurable productivity increases

MONTH 11 — Build Your AI Culture

Now align the team.

You will:

- train all employees
- assign AI responsibilities
- eliminate outdated processes
- set company-wide standards
- implement AI ethics & compliance

Deliverables:

✓ a trained, AI-enabled workforce

- ✓ an internal AI policy
- ✓ updated job descriptions
- ✓ quarterly AI training plan

MONTH 12 — Strategic Future-Proofing

Now that your business is AI-powered, you set yourself up for the next 5 years.

You will create:

- an annual AI budget
- a tool maintenance plan
- a quarterly workflow update schedule
- a tool evaluation calendar
- a growth acceleration plan

Deliverables:

- ✓ annual AI roadmap
- ✓ annual tool budget
- ✓ 12-month forecasting model
- ✓ competitive strategy

By the end of Year 1, you will have:

- an AI-powered business
- modern workflows
- a trained team
- a scalable tech stack
- faster operations
- stronger marketing
- better customer service
- reduced overhead
- increased revenue
- competitive insulation

SECTION 6 — The 12-Month AI Roadmap (At a Glance)

Q1 — Learn & Stabilize (Months 1–3)

- ✓ 1 workflow
- ✓ 1 tool per category
- ✓ marketing basics
- ✓ customer response automation

Q2 — Scale & Systemize (Months 4–6)

- ✓ operations automation
- ✓ marketing engine
- ✓ CRM automation
- ✓ sales automation

Q3 — Optimize & Document (Months 7–9)

- ✓ refine systems
- ✓ build SOPs
- ✓ analytics dashboards
- ✓ forecasting

Q4 — Accelerate & Future-Proof (Months 10–12)

- ✓ AI agents
- ✓ team-wide adoption
- ✓ annual plan
- ✓ competitive advantage

MOVING FORWARD

You Now Have the Clear Path Forward Most Businesses Will Never See

The next 12 months will determine:

- your competitiveness
- your profitability

- your customer experience
- your time freedom
- your market leadership
- your resilience

Most businesses won't adopt AI this year.
Some will experiment but quit early.
Only a small percentage will follow a structured roadmap like this.

Those are the businesses that will dominate the next decade.

You now have the roadmap.
The only thing left is action.

A Moment for Reflection

We are close to the summit. Before you take another step or implement more tools, automate more workflows, or scale what you've started, I want you to pause for just a moment.

Because while so much is changing...

Some things should not.

In fact, they matter more now than ever.

Trust

Technology can move fast. Faster than we sometimes fully understand.

But trust is still built the same way it always has been:

- doing what you say you'll do

- showing up consistently
- treating people fairly

AI can help you respond faster.
It can help you communicate better.

But it cannot replace trust.

And if you ever sacrifice trust for speed, convenience, or short-term gain...

You will lose something that no system can rebuild for you.

Craft

There is still a right way to do things.

Whether you run a service business, a healthcare organization, a retail operation, or a professional firm, your work matters.

The quality of what you deliver matters.

AI can help you:

- draft faster
- analyze faster
- produce more

But it should never lower your standards.

If anything, it should raise them.

Because when the tools become more powerful, the responsibility to use them well becomes greater.

Your craft, your attention to detail, your standards, your pride in the work is all still yours.

Protect it.

Relationships

At the end of the day, business is still human.

Customers don't remember your systems.
They remember how you made them feel.

Your team doesn't follow tools.
They follow leadership.

AI can enhance communication.
It can improve responsiveness.
It can remove friction.

But it should never remove connection.

If anything, it should give you back the time and space to be more present:

- with your customers
- with your team
- with the people who trust you

Because relationships are still the foundation everything else is built on.

Leadership

This is where it all comes together.

AI is powerful.

It can influence decisions, shape communication, and drive outcomes at a scale we've never seen before.

And that's exactly why leadership matters more now—not less.

Because someone must decide:

- where it's used
- how it's used
- and where it shouldn't be used

Someone has to set the standard and protect the culture.

Someone must ensure that efficiency never comes at the cost of integrity.

That someone... is you.

The Role of Humans in an AI-Powered World

AI is not the decision-maker.

It's not the owner.
It's not the leader.
It's not accountable.

You are.

Your judgment matters.
Your values matter.
Your standards matter.

More than ever.

Because without clear thinking, strong principles, and intentional leadership, powerful tools can be misused.

Shortcuts can be taken.
Corners can be cut.
Trust can be broken.

And when that happens, the damage goes far beyond business metrics.

This Matters More Than Profit

Yes, AI can help you:

- grow faster
- operate more efficiently
- increase margins

And those things matter.

But not at the expense of:

- your reputation
- your integrity
- your responsibility to your customers and your team

Because in the long run, the businesses that win aren't just the most efficient.

They're the most trusted.

They're the most consistent.

They're the ones that use powerful tools responsibly.

As You Move Forward

Build the systems.
Use the tools.
Embrace the change.

But don't lose sight of what makes a business great.

Stay grounded in:

- trust
- craftsmanship
- relationships
- leadership

Because AI will amplify whatever foundation you already have.

If that foundation is strong...

You will grow faster than you thought possible.

If it's not...

The cracks will show just as quickly.

You are not just building a more efficient business.

You are building a better one.

One that is:

- smarter
- faster
- more capable

But also, more human, where it matters most.

Use AI to move faster.
But lead in a way that makes it worth it.

That's how you win.

Not just now...

But for the long run.

CHAPTER 17: The Future of AI for Small Business

What the Next 3–10 Years Will Look Like — And How to Make Sure You're on the Winning Side of History

VALUE PROPOSITION

The Biggest Change in Business Since the Industrial Revolution Has Already Started and Most small businesses Don't See It Yet

Artificial intelligence is moving faster than any technology in human history.
It's not slowing down.
It's not stabilizing.
It's not plateauing.

AI today is the **flip phone era**.
AI next year is the **iPhone era**.
AI in five years is the **self-driving, personalized, automated, self-learning business ecosystem** era.

Small businesses will feel this shift more than anyone.

The good news?
Small businesses also have the most to gain.

The businesses that start now will dominate their markets.
The businesses that wait will be invisible.

This chapter prepares you for the next decade — the opportunities, the disruptions, the new workflows, the new customers, and the entirely new business landscape that AI is creating.

You are not reading a prediction chapter.
You are reading a **blueprint for the future**.

SECTION 1 — The 12 Transformations That Will Reshape Small Business Forever

AI won't just change business.
It will **redefine what a business is.**

These are the 12 biggest shifts already underway:

1. AI Will Become the New Employee

(Every Business Will Have a Digital Workforce)

AI won't assist humans — humans will assist AI.

Small businesses will soon have:

- AI receptionists
- AI schedulers
- AI marketers
- AI sales reps
- AI bookkeepers
- AI HR assistants
- AI compliance monitors
- AI inventory controllers

Your team will be smaller — but far more productive.

This will redefine hiring, payroll, training, and workforce structure.

2. Websites Will Be Replaced by AI Agents

Instead of clicking buttons, customers will have full conversations with your business:

"Book me for Thursday at 3."
"Which service is right for me?"
"Can you send a quote?"
"Do you have availability next week?"
"Show me the best options."

Your current website will look like a fax machine compared to what's coming.

3. Marketing Will Become Fully Automated

AI will:

- create ads
- test variations
- update creative
- rewrite poor-performing posts
- optimize daily
- adjust based on customer behavior
- retarget instantly
- forecast trends
- maintain brand voice consistently

AI won't support marketing.
AI **is** the marketing department.

4. Sales Will Be 24/7 — Powered By AI

Lead responds at 2:13 AM?
No problem.

AI will:

- qualify
- overcome objections
- schedule the call
- send follow-up
- deliver documents
- score the lead
- notify the human rep if needed

Speed-to-lead becomes instantaneous — permanently.

5. Operations Will Become Fully Predictive

AI will know what's going wrong before you do.

It will:

- predict scheduling conflicts
- identify staffing shortages
- detect customer dissatisfaction
- flag invoices at risk
- forecast inventory needs
- automate reordering
- optimize daily workflow

Small business operations will run like Fortune 500 companies.

6. Customer Expectations Will Skyrocket

Customers will expect:

- instant responses
- perfect accuracy
- personalized messaging
- seamless experiences
- proactive service

Businesses using AI will deliver it.
Businesses without AI won't even come close.

This creates a **permanent competitive gap**.

7. AI Will Become a Mandatory Compliance Layer

AI will:

- monitor documentation
- verify steps
- flag errors
- check regulatory requirements
- track employee actions
- reduce risk

Most industries — healthcare, finance, legal, construction, accounting — will require AI for compliance.

Not optional.
Mandatory.

8. The Cost of AI Will Plummet (While Power Increases 10x)

In 3–5 years:

- AI tools will drop 75% in cost
- AI features will be bundled into every platform
- computing will become cheaper
- competition will drive innovation
- agents will evolve into full "digital employees"

You'll get 10x more capability for a fraction of the price.

9. AI Will Create Super-Solo Entrepreneurs ("1-Person Empires")

With AI:

- one person can do the work of 5–15 people
- marketing is automated
- admin is automated
- lead gen is automated
- sales is assisted
- finance is AI-driven
- content is AI-generated

The "solo founder" becomes a legitimate competitor to multimember companies.

10. Small Teams Will Outperform Large Ones

This is the great equalizer.

A 3-person company with AI will outperform a 20-person company without it.

Productivity becomes exponential.
Payroll becomes leaner.
Overhead becomes lighter.
Quality becomes more consistent.
Customer experience becomes flawless.

Small becomes strong.

11. Traditional Job Roles Will Disappear (and New Ones Will Replace Them)

Jobs disappearing:

- receptionist
- dispatcher
- basic support agent
- junior admin
- junior marketing assistant
- entry-level analyst
- basic bookkeeper
- appointment setter

Jobs appearing:

- AI workflow designer
- AI prompt strategist
- AI operations manager
- AI content coordinator
- AI customer experience architect
- AI ethics/review auditor
- hybrid AI-support roles

Your hiring strategy must evolve.

12. Small Businesses Will Adopt AI Faster Than Corporations

Corporations:

- move slowly
- have committees
- require approvals
- deal with legacy systems
- face bureaucratic blockers

Small businesses:

- move fast
- adapt quickly

- test immediately
- pivot instantly

Small business becomes the **innovation leader**, not the follower.

SECTION 2 — The 5 AI Predictions Every Business Must Prepare For

These aren't guesses.
These are inevitabilities.

Prediction #1: Every Business Will Need an AI Policy

Customers will demand it.
Vendors will require it.
Employees will depend on it.

Policies will cover:

- data privacy
- human oversight
- tool selection
- staff training
- compliance
- ethical guardrails

You'll have one.
It will be updated frequently.
It will be non-negotiable.

Prediction #2: AI Agents Will Replace 50–60% of Frontline Interactions

Not humans — **interactions**.

Customers will speak with AI first.

Humans will handle the exceptions.

This is already happening in:

- home services
- dentistry
- real estate
- medical offices
- retail
- hospitality
- accounting
- insurance
- salons & spas

Speed wins the customer — and AI is always the fastest.

Prediction #3: Competition Will Become Brutal for AI Holdouts

In 3–5 years, businesses without AI will:

- respond slower
- market weaker
- operate inefficiently
- spend more on labor
- lose customers
- get outcompeted on pricing
- experience burnout

Some won't survive.

This isn't fear.
It's math.

Prediction #4: Customer Experience Will Become Fully Personalized

AI will tailor:

- recommendations
- offers
- follow-ups
- content
- timing
- messaging
- service levels

Every customer gets a “white glove” experience — automatically.

Customers will feel understood by your business in ways previously impossible.

Prediction #5: The SMB Winners Will Be the Ones Who Build AI Literacy Now

The biggest shift won’t be tools.

It will be people.

The SMB owner who understands:

- what AI can do
- how it behaves
- how to prompt it
- how to deploy it
- how to measure ROI
- how to lead an AI-enabled workforce

...will be unstoppable.

SECTION 3 — What Small Businesses Should Do *Right Now* to Prepare for the Future

This is your future-proofing checklist.

1. Build your AI tool belt (5 core tools)

- ✓ ChatGPT or Claude
- ✓ AI CRM (HubSpot or GHL)
- ✓ Customer response AI (Tidio/Air.ai)
- ✓ Canva AI
- ✓ Zapier

2. Build an AI culture

Train employees every month.
Reward AI adoption.
Make improvement normal.

3. Document everything using AI

Use AI to generate SOPs, playbooks, policies, and training.

4. Build one AI agent every quarter

- AI receptionist
- AI marketing assistant
- AI sales assistant
- AI onboarding assistant

5. Make AI part of your strategic planning

Ask:
"How can AI improve this?"
at every decision.

6. Shift your hiring strategy

Look for:

- Adaptability
- Curiosity
- tech comfort
- willingness to learn

These traits matter more than experience.

7. Measure AI performance monthly

Track:

- time saved
- errors avoided
- leads captured
- revenue increases
- customer satisfaction

AI must be tied to outcomes.

SECTION 4 — The Biggest Lie About AI and Small Business

People think AI will replace small business owners.

It won't.

AI will make small-business owners **more important than ever**.

AI cannot replace:

- leadership
- vision

- personality
- relationships
- creativity
- empathy
- community
- local reputation

AI does the heavy lifting.
Humans do the leading.

The future is **AI-powered, human-led** business.

MOVING FORWARD

The Future Belongs to the Small Businesses Who Adapt — Starting Now

You now know what the next decade will look like.
This isn't speculation.
This is direction.

For small businesses, AI is:

- the great equalizer
- the great accelerator
- the great competitive advantage

The future isn't scary.
It's promising.
It's exciting.
It's full of opportunity.

And you are early — but only if you act now.

Your future competitors are already building AI-enabled businesses.
Your future customers are already expecting AI-level execution.
Your future workforce is already being shaped by AI.

The question isn't whether AI will change your business.
The question is:

Will you lead the change or be forced into it?

You now have the tools, the roadmaps, and the strategies.
You are no longer guessing.
You are no longer in the dark.
You are no longer behind.

Turn the page.

Your AI-powered future starts now.

WORKSHEETS & CHECKLISTS

CHAPTER 1 WORKSHEET & CHECKLIST

The Great Business Divide: AI Adopters vs. AI Holdouts

Quick Summary

This chapter proves the competitive gap is already here: businesses using AI are pulling ahead in speed, efficiency, marketing, sales, and customer experience — while holdouts fall behind rapidly.

YOUR CHAPTER 1 SELF-ASSESSMENT

1. Where does your business currently stand?

(Choose one)

☐ AI Adopter

☐ Early Tester

☐ Skeptical but Curious

☐ Overwhelmed

☐ AI Holdout

2. Identify Your Current Risks (check all that apply):

☐ Slow response times

☐ Inconsistent marketing

☐ Missed leads

☐ Repetitive manual tasks

☐ High labor costs

☐ Customer complaints

☐ Falling behind competitors

3. Identify Your Opportunities:

☐ Faster customer communication

☐ Better follow-up

☐ Automated scheduling

☐ Automated marketing

☐ Better decision-making

☐ Streamlined operations

☐ Reduced errors

☐ Team support & productivity

☐ Higher profitability

CHAPTER 1 IMPLEMENTATION CHECKLIST

☐ Identify whether you're currently an adopter or holdout

☐ Review where competitors use AI

☐ Identify 1–2 areas where you feel "behind"

☐ Identify 1–2 areas where AI could instantly help

☐ Re-read the Case Studies (Success + Failure)

☐ Commit to crossing the divide

CHAPTER 1 ACTION PLAN

My business is currently on the ______ side of the AI divide.
Our biggest competitive threats are:

______________________________.

Our immediate AI opportunities are:

______________________________.

Our first step toward becoming an AI-powered business will be:

____

____________________________.

CHAPTER 2 WORKSHEET & CHECKLIST

Clearing the Fog: What AI Actually Is (and Isn't)

Quick Summary

This chapter removes confusion by explaining AI in plain English and showing what AI can (and cannot) do for small businesses.

YOUR CHAPTER 2 UNDERSTANDING CHECK

1. Rate your clarity on AI before reading this chapter:

1 = Confused, 10 = Confident
Before: ___ / 10
After: ___ / 10

2. What AI actually is (write in your own words):

AI is:

____

___________________________.

3. What AI is NOT (check all misunderstood areas you had):

☐ Complicated

☐ Expensive

☐ A replacement for employees

☐ Only for big companies

☐ A trend

☐ Dangerous by default

☐ Something that requires a tech background

---

4. AI Capabilities Your Business Could Use Immediately

☐ Understanding information (reading emails/PDFs)

☐ Generating information (writing content)

☐ Analyzing information (financials, patterns)

☐ Acting on information (automated tasks)

Check the top two that matter most:

#1 ____________________
#2 ____________________

CHAPTER 2 IMPLEMENTATION CHECKLIST

☐ Understand the difference between Automation, ML, and AI

☐ Recognize at least 4 capabilities AI can bring to your business

☐ Write a personal definition of AI

☐ Identify 2–3 workflows where AI can help

☐ Eliminate misconceptions

CHAPTER 2 ACTION PLAN

AI capability I will use first:

Why this matters for my business:

CHAPTER 3 WORKSHEET & CHECKLIST

The 10 Biggest Myths Keeping SMBs from Using AI

Quick Summary

Your biggest barriers are psychological, not technological. This chapter destroys the myths that keep 90% of SMBs stuck.

MYTH IDENTIFICATION WORKSHEET

1. Which myths did YOU believe (check all that apply):

☐ "AI is too expensive."

☐ "AI will replace my team."

☐ "I'm not tech-savvy enough."

☐ "AI makes business impersonal."

☐ "My customers don't care about AI."

☐ "AI will embarrass my business."

☐ "AI takes too long to set up."

☐ "AI will cause legal/compliance issues."

☐ "AI is a fad."

☐ "AI will destroy my business if I use it wrong."

---

2. List the top 3 myths holding YOU back:

1. ______________________
2. ______________________
3. ______________________

3. The truth that replaces your biggest myth:

My biggest myth was:

____

The reality is:

CHAPTER 3 IMPLEMENTATION CHECKLIST

☐ Identify myths you believed

☐ Replace myths with facts

☐ Discuss this chapter with your team

☐ Commit to learning (not fearing) AI

☐ Document which myths slowed you down

CHAPTER 3 ACTION PLAN

The myth that held me back the most:

How I will overcome it:

CHAPTER 4 WORKSHEET & CHECKLIST

The Overwhelm Problem — "I Don't Know Where to Start"

Quick Summary

Overwhelm is the #1 reason small businesses avoid AI. This chapter gives practical clarity and removes confusion.

THE OVERWHELM DIAGNOSIS WORKSHEET

1. What's overwhelming you? (Check all that apply)

☐ Too many tools

☐ Too many opinions

☐ Too much hype

☐ Fear of choosing wrong

☐ Fear of breaking something

☐ Feeling "behind"

☐ Not knowing where to start

☐ No time to test

---

2. List your top 3 "overwhelm" sources:

1. ______________________________

2. ______________________________

3. ______________________________

3. Apply The One Workflow Rule

Which ONE workflow meets all three criteria?

✓ High friction
✓ High repetition
✓ High cost (time or money)

My One Workflow is:

____

____

CHAPTER 4 IMPLEMENTATION CHECKLIST

☐ Identify overwhelm sources

☐ Pick ONE workflow

☐ Ignore all other tools for now

☐ Commit to 30 days of focus

☐ Set expectations with team

CHAPTER 4 ACTION PLAN

The workflow I will fix in Month 1:

____

Why this workflow was chosen:

The result I want:

CHAPTER 5 WORKSHEET & CHECKLIST

Your First 30 Days with AI — The Single Workflow AI Method

Quick Summary

This chapter gives you a step-by-step, stress-free method to implement AI in 30 days. Simple, focused, and guaranteed to work.

THE 30-DAY BLUEPRINT WORKSHEET

WEEK 1 — DIAGNOSE THE BOTTLENECK

Identify your biggest bottleneck:

✓ Time wasted on:

____

___

✓ Money lost due to:
____

_

✓ Customer frustration caused by:
____

WEEK 2 — PICK THE RIGHT TOOL (One Only)

The tool I will use is:

____

Why I chose it:

____

____

WEEK 3 — IMPLEMENT (One Automation Only)

My ONE automation is:

____

Success looks like:

____

____

WEEK 4 — MEASURE ROI

Time saved this month:

________ hours

Money saved or earned:

$__________

Customer improvements:

CHAPTER 5 IMPLEMENTATION CHECKLIST

- ☐ Identified bottleneck
- ☐ Chose workflow
- ☐ Selected ONE tool
- ☐ Built ONE automation
- ☐ Measured results
- ☐ Documented learnings
- ☐ Prepared for next workflow

CHAPTER 5 ACTION PLAN

My Month 1 AI win will be:

And after that, I will automate:

✔ **Let's move on to the NEXT SET.**

CHAPTER 6 WORKSHEET & CHECKLIST

The SMB AI Tech Stack — The 10 Essential Tools Every Business Needs

Quick Summary

Your tech stack determines your speed, consistency, accuracy, and competitiveness. You don't need 30 tools — you need the right 5–7.

TECH STACK SELF-ASSESSMENT

1. Rate your current tech stack strength:

1 = outdated, 10 = strong
Score: ___ / 10

2. Which of these do you already have? (Check all that apply)

AI Writing Tools

☐ ChatGPT

☐ Claude

☐ Gemini

☐ Perplexity

AI Customer Response Tools

☐ Tidio

☐ Intercom

☐ Air.ai

☐ Smith.ai

AI CRM & Sales Automation

☐ HubSpot

☐ GoHighLevel

☐ Close CRM

☐ Salesforce SMB

AI Marketing & Design Tools

☐ Canva AI

☐ Jasper

☐ Adobe Firefly

Workflow Automation Engines

☐ Zapier

☐ Make

Scheduling Tools

☐ Calendly

☐ TidyCal

Finance Tools

☐ QuickBooks AI

☐ Wave AI

☐ Xero AI

HR & Hiring Tools

☐ BambooHR AI

☐ Rippling AI

☐ Homerun AI

SOP & Knowledge Management

☐ Notion AI

☐ Guru AI

Industry-Specific AI Tools

☐ ________________________

3. What tools are missing from your stack?

My missing categories are:

The top missing tool I need most is:

____

____

CHAPTER 6 IMPLEMENTATION CHECKLIST

☐ Identify your 1–2 highest ROI categories

☐ Select ONE tool per category

☐ Eliminate redundant or unused subscriptions

☐ Document your current stack

☐ Train one "AI Champion" on each tool

☐ Integrate tools using Zapier or Make

☐ Set up basic safeguards (permissions, roles, access limits)

CHAPTER 6 ACTION PLAN

The next tool I will implement is:

Because it will help with:

And my team member responsible is:

CHAPTER 7 WORKSHEET & CHECKLIST

AI for Marketing & Lead Generation — Build a 24/7 Content & Lead Engine

Quick Summary

Marketing is the top business use case for AI. This chapter gives you strategy, content, distribution, follow-up, and optimization in a unified system.

MARKETING DIAGNOSTIC WORKSHEET

1. What's broken or inconsistent in your marketing? (Check all)

☐ Posting inconsistency

☐ Weak branding

☐ Not enough content

☐ Slow responses

☐ Not enough reviews

☐ No follow-up system

☐ No clear marketing strategy

☐ No analytics or tracking

☐ Low engagement

☐ Weak ad performance

2. Identify your top marketing goal:

Pick ONE:

☐ More leads

☐ More content

☐ More visibility

☐ Higher engagement

☐ Better branding

☐ Better follow-up

☐ Better reviews

☐ More website traffic

My #1 goal is:

____

____

3. Build Your AI Marketing Toolkit (Choose One per Category)

AI Writing & Strategy Engine:

☐ ChatGPT ☐ Claude ☐ Jasper ☐ Other: _______

Design Engine:

☐ Canva AI ☐ Adobe Firefly

Video/Reels Engine:

☐ Vidyo.ai ☐ OpusClip ☐ Descript

CRM/Follow-Up Engine:

☐ HubSpot ☐ GoHighLevel ☐ Close

Website Chat/Lead Capture:

☐ Tidio ☐ Intercom ☐ Air.ai

Automation Layer:

☐ Zapier ☐ Make

CHAPTER 7 IMPLEMENTATION CHECKLIST

☐ Create your brand voice file

☐ Build a year-long content strategy

☐ Generate 30 days of content

☐ Set up an AI follow-up engine

☐ Set up an AI review system

☐ Connect everything through automation

☐ Build your analytic dashboard

CHAPTER 7 ACTION PLAN

My #1 Marketing Focus:

____

AI tools I will implement first:

____

What success looks like in 30 days:

____

____

____

CHAPTER 8 WORKSHEET & CHECKLIST

AI for Sales & Follow-Up — Close More Deals, Automatically

Quick Summary

AI doesn't replace salespeople — it eliminates the delays, inconsistency, and lost leads that kill revenue. This chapter helps build a modern AI-powered sales engine.

SALES DIAGNOSTIC WORKSHEET

1. Identify the gaps in your sales process (check all):

☐ Slow lead response

☐ No follow-up system

☐ Leads slipping through cracks

☐ Little or no lead scoring

☐ Manual pipeline updates

☐ No automated emails/SMS

☐ Poor CRM usage

☐ Inconsistent sales scripts

☐ No reporting or tracking

2. Rate your current sales system strength:

1 = weak, 10 = strong
Score: ___ / 10

3. Your Sales Stack Setup

CRM:
☐ HubSpot ☐ GoHighLevel ☐ Close ☐ Salesforce SMB

AI Lead Response:
☐ Tidio ☐ Intercom ☐ Air.ai

Follow-Up Automation:
☐ HubSpot Sequences
☐ GoHighLevel Workflows
☐ Zapier

CHAPTER 8 IMPLEMENTATION CHECKLIST

☐ Build instant email/SMS replies

☐ Build a 7-day follow-up automation

☐ Create AI sales scripts

☐ Implement lead scoring

☐ Build proposal/estimate templates

☐ Create pipeline stages

☐ Build weekly sales reporting

☐ Set up automated reminders

☐ Document the entire system

CHAPTER 8 ACTION PLAN

The biggest leak in my sales process is:

____

The AI fix I'll implement this month:

____

Success will look like:

____

____

____

CHAPTER 9 WORKSHEET & CHECKLIST

AI for Customer Service & Appointment Management

Quick Summary

AI can instantly reduce customer frustration, speed up response times, and handle scheduling, FAQ, and support tasks on autopilot.

CUSTOMER SERVICE WORKFLOW WORKSHEET

1. Current challenges (check all):

☐ Too many telephone calls

☐ Slow replies

☐ Missed messages

☐ Repetitive questions

☐ Scheduling chaos

☐ No automated reminders

☐ No after-hours support

☐ High no-show rates

☐ Inconsistent answers from staff

2. Your Ideal AI Customer Service Engine

AI Chat:
☐ Tidio ☐ Intercom ☐ Smith.ai

AI Voice:
☐ Air.ai

Scheduling:
☐ Calendly ☐ TidyCal

Reminders:
☐ GoHighLevel
☐ Zapier Automation

CHAPTER 9 IMPLEMENTATION CHECKLIST

☐ Build your FAQ database

☐ Train your AI receptionist/chat

☐ Set up scheduling automations

☐ Set up reminders + confirmations

☐ Build after-hours response system

☐ Add AI to website + Google Profile

☐ Set up missed-call text back

☐ Test with 20–30 customers

☐ Monitor accuracy weekly

CHAPTER 9 ACTION PLAN

The biggest customer service bottleneck is:

___

The AI solution I'll add first:

____

The expected impact:

____

____

____

CHAPTER 10 WORKSHEET & CHECKLIST

AI for Operations & Admin — Run Leaner, Faster, and Error-Free

Quick Summary

Operations is where small businesses lose the most time. AI creates consistency, eliminates repetition, and frees your team to focus on high-value work.

OPERATIONS DIAGNOSTIC WORKSHEET

1. Which areas of operations are slowing you down? (Check all)

☐ Scheduling

☐ Billing / invoicing

☐ Data entry

☐ Inventory

☐ SOP documentation

☐ Internal communication

☐ Reporting

☐ Onboarding

☐ Task management

☐ Compliance

☐ Vendor/partner coordination

☐ Project management

2. Your AI Operations Stack

Automation:
☐ Zapier ☐ Make

Documentation:
☐ Notion AI ☐ Guru AI

Finance:
☐ QuickBooks AI ☐ Wave AI ☐ Xero AI

Scheduling:
☐ Calendly ☐ TidyCal

Internal Communication:
☐ Slack AI
☐ Teams AI

CHAPTER 10 IMPLEMENTATION CHECKLIST

☐ Document your top 3 admin bottlenecks

☐ Automate 1 repetitive workflow

☐ Build internal SOPs with AI

☐ Automate weekly reporting

☐ Automate onboarding tasks

☐ Create automated reminders

☐ Eliminate manual data entry

☐ Connect accounting tools

☐ Train team on new workflows

CHAPTER 10 ACTION PLAN

The operational workflow I will automate first is:

____

The tool I will use:

____

The time savings I expect:

____

____

____

CHAPTER 11 WORKSHEET & CHECKLIST

AI for HR & Hiring — Build a Stronger, Faster, More Efficient Team

Quick Summary

AI transforms hiring from guesswork to a streamlined, data-driven, bias-reduced system. Small businesses finally get recruiting power that used to belong only to large companies.

HR & HIRING DIAGNOSTIC WORKSHEET

1. What HR pain points is your business dealing with? (Check all)

☐ Too few applicants

☐ Too many unqualified applicants

☐ Hard-to-write job descriptions

☐ Slow communication with candidates

☐ Difficult screening

☐ No structured interview process

☐ High turnover

☐ Slow onboarding

☐ Poor documentation

- ☐ Compliance concerns
- ☐ Inconsistent training

2. AI Tools You Already Have in HR:

Hiring:
☐ BambooHR AI ☐ Rippling AI ☐ Homerun AI

Screening:
☐ ChatGPT/Claude (for resume review)
☐ Applicant Tracking System with AI

Onboarding:
☐ Notion AI
☐ Guru
☐ Internal checklist in CRM

CHAPTER 11 IMPLEMENTATION CHECKLIST

- ☐ Generate job descriptions using AI
- ☐ Create standardized screening criteria
- ☐ Use AI to score resumes
- ☐ Build automated follow-up to candidates
- ☐ Draft interview questions with AI
- ☐ Create onboarding documents with AI
- ☐ Build an onboarding checklist
- ☐ Automate new-hire paperwork

☐ Design an AI-assisted training library

☐ Document your entire hiring workflow

CHAPTER 11 ACTION PLAN

My next role to hire:

____

The AI tools I'll use in hiring:

____

What success looks like:

____

____

____

CHAPTER 12 WORKSHEET & CHECKLIST

AI for Finance & Accounting — Cleaner Books, Faster Decisions, No More Guessing

Quick Summary

AI helps eliminate financial blind spots, reduce manual bookkeeping, and give business owners a real-time picture of their cash flow, expenses, and profitability.

FINANCE DIAGNOSTIC WORKSHEET

1. Which areas of your financial system need the most help? (Check all)

☐ Late invoicing

☐ Slow bookkeeping

☐ Unreliable cash flow forecasts

☐ Mismatched transactions

☐ Expense categorization

☐ Overwhelmed accountant

☐ Slow reporting

☐ No financial dashboard

☐ Difficulty tracking profitability by job/client

☐ Tax-season panic

☐ No reminders for unpaid invoices

2. AI Tools in Your Finance Stack

Accounting:
☐ QuickBooks AI ☐ Wave AI ☐ Xero AI

Analytics:

☐ Microsoft Excel AI

☐ Google Sheets AI

☐ ChatGPT data analysis

Invoicing / Billing:

☐ Stripe

☐ QuickBooks

☐ FreshBooks

☐ Zoho

CHAPTER 12 IMPLEMENTATION CHECKLIST

☐ Connect your accounting system to AI features

☐ Automate expense categorization

☐ Set up invoice reminders

☐ Build an AI monthly financial summary

☐ Create a cash flow projection workflow

☐ Build a profitability dashboard

☐ Document financial SOPs

☐ Automate bill-pay reminders

☐ Implement "financial review day" on your calendar

CHAPTER 12 ACTION PLAN

The biggest financial bottleneck I have is:

The AI tool that will fix it is:

I expect to save/earn:

CHAPTER 13 WORKSHEET & CHECKLIST

The AI-Enabled Workforce — Build a Future-Proof Team That Thrives, Not Breaks

Quick Summary

AI doesn't replace people — it elevates them. This chapter helps identify where to upskill, how to build an AI-positive culture, and how to align roles with new AI capabilities.

AI WORKFORCE DIAGNOSTIC WORKSHEET

1. Assess your current team readiness (score 1–10)

- Willingness to learn AI: ____
- Comfort with new tools: ____

- Repetitiveness of tasks they perform: ____
- Ability to adapt to change: ____
- Documentation quality: ____
- Tech literacy: ____
- Initiative & curiosity: ____

2. Identify AI Opportunities for Each Role

Fill in for your team:

Employee / Role	Repetitive Tasks They Do	What AI Could Automate	Training Needed
________	________________	________________	__________
________	________________	________________	__________
________	________________	________________	__________

3. Your AI Upskilling Plan

Rate priority for each skill (1–5):

- Basic prompt writing ☐1 ☐2 ☐3 ☐4 ☐5
- Using AI for communication ☐1 ☐2 ☐3 ☐4 ☐5
- Using AI for task automation ☐1 ☐2 ☐3 ☐4 ☐5
- Using AI for reporting/data ☐1 ☐2 ☐3 ☐4 ☐5
- Using AI for documentation ☐1 ☐2 ☐3 ☐4 ☐5

CHAPTER 13 IMPLEMENTATION CHECKLIST

☐ Identify your first "AI Champion"

☐ Train team on basic AI prompting

☐ Document each role's AI opportunities

☐ Implement role-specific automations

☐ Create SOPs using AI

☐ Build an internal "AI resource hub"

☐ Review new workflows every 30 days

☐ Hold quarterly AI upskilling workshops

CHAPTER 13 ACTION PLAN

The first role I will augment with AI:

____

Automation I will create:

____

Training my team needs next:

____

____

CHAPTER 14 WORKSHEET & CHECKLIST

AI SAFETY, COMPLIANCE & TRUST

Protect your business, your customers, and your reputation.

Quick Summary

AI is powerful — but only when used responsibly. This chapter helps you safeguard data, create policies, avoid legal risks, and build customer trust.

AI SAFETY DIAGNOSTIC WORKSHEET

1. What risks apply to your business? (Check all)

Data Risks

☐ Sensitive customer information

☐ Payment information

☐ Medical or health information

☐ Location data

☐ Employee data

☐ Confidential client files

Operational Risks

☐ Incorrect answers from AI

☐ Miscommunications

☐ Workflow failures

☐ Automations triggering incorrectly

Compliance Risks

☐ HIPAA

☐ OSHA

☐ PCI / credit card security

☐ State-level data privacy rules

☐ Employment law

YOUR AI TRUST & COMPLIANCE SCORECARD

Rate yourself from 1–5:

Category	Score (1–5)
Customer data protection	___
Employee data protection	___
Accuracy safeguards	___
Human review processes	___
Documentation policies	___

Category	Score (1–5)
Automation safety	___
Customer transparency	___
Vendor vetting	___

CHAPTER 14 IMPLEMENTATION CHECKLIST

AI Policy

- ☐ Create a formal AI usage policy
- ☐ Define what data can and cannot be used
- ☐ Define when human review is required
- ☐ Define approval process for new tools

Tool Safety

☐ Enable privacy settings inside AI tools

☐ Use business-tier accounts when possible

☐ Confirm where your data is stored

☐ Confirm vendors do NOT use your data to train models

Data Protection

☐ Avoid entering sensitive data in public models

☐ Encrypt stored data

☐ Train staff on safe AI practices

☐ Use masked or redacted data when possible

Workflow Safety

☐ Test every automation 10–20 times before going live

☐ Build "stop conditions" to prevent runaway automations

☐ Keep humans involved in all judgment-based decisions

CHAPTER 14 ACTION PLAN

The biggest safety risk in my business is:

____

The first safeguard I will implement is:

____

Tools/Policies I will review this month:

____

____

____

____

CHAPTER 15 WORKSHEET & CHECKLIST

CHOOSING THE RIGHT TOOLS WITHOUT GETTING SCAMMED

How to avoid garbage software, overpriced tools, fake gurus, and bad investments.

Quick Summary

Most AI tools fail small businesses because they are:
- overly complicated
- overpriced
- unnecessary
- poorly supported
- built by unknown companies

This chapter teaches you what to avoid — and what to buy.

TOOL VETTING WORKSHEET

1. Purpose Check

Before buying ANY tool, answer:

What exact workflow will this tool automate?

If I don't automate this, what is the cost of doing nothing?

Which employee/job does this tool reduce or support?

---

2. Tool Quality Checklist

Rate each from 1–5:

- Company reputation: ____
- Number of real reviews: ____
- Age of the company: ____
- Security/privacy transparency: ____
- Clarity of pricing: ____
- Support responsiveness: ____
- Ease of setup: ____
- Data protection: ____
- Integration options (Zapier, API): ____
- Does it solve a *real* problem? (not a shiny object): ____

THE 6-QUESTION "SCAM FILTER"

If ANY of these are "YES," walk away.

☐ Is the pricing unclear?

☐ Is there no real company address or owner listed?

☐ Are the reviews suspicious or generic?

☐ Are the features too vague to understand?

☐ Is the tool new with no history?

☐ Is there pressure to "buy today only"?

CHAPTER 15 IMPLEMENTATION CHECKLIST

☐ Identify your exact workflow before buying any tool

☐ Compare only 2–3 tools

☐ Read real reviews (G2, Capterra, Reddit)

☐ Test the free trial

☐ Check integration requirements

☐ Set a max budget ($50–$200/mo)

☐ Train one person before rolling out

☐ Document the tool's purpose

☐ Cancel tools that don't produce ROI in 30 days

CHAPTER 15 ACTION PLAN

My next AI tool purchase will be for this workflow:

The tools I'm comparing:

My ROI expectation:

CHAPTER 16 WORKSHEET & CHECKLIST

YOUR 12-MONTH AI TRANSFORMATION ROADMAP

The complete, step-by-step plan that takes your business from manual › automated › scalable.

Quick Summary

This chapter turns everything in the book into a practical, 12-month execution plan.

12-MONTH READINESS SELF-ASSESSMENT

Rate yourself 1–5:

Category	Score
AI mindset readiness	___
Workflow clarity	___
Team willingness	___
Budget readiness	___
Tool understanding	___
Automation comfort	___
Documentation quality	___
Customer communication readiness	___

THE 12-MONTH ROADMAP PLANNER

Quarter 1 — Foundation

- ☐ Identify top 3 workflows
- ☐ Pick initial tools
- ☐ Train AI champion
- ☐ Implement first 2 automations
- ☐ Build prompt library
- ☐ Create AI policy

Quarter 2 — Expansion

☐ Expand automations into customer service

☐ Build AI-driven marketing engine

☐ Build follow-up sequences

☐ Train remaining team members

☐ Document new SOPs

Quarter 3 — Optimization

☐ Add financial automations

☐ Add HR automation

☐ Add reporting dashboard

☐ Upgrade prompts, templates, automations

☐ Remove unused tools

Quarter 4 — Scaling

☐ Automate high-impact departments

☐ Build industry-specific AI workflows

☐ Add voice AI or AI agents

☐ Formalize training library

☐ Create AI-powered strategic plan for next year

CHAPTER 16 IMPLEMENTATION CHECKLIST

☐ Choose one workflow per quarter

☐ Choose one tool per workflow

☐ Implement with human review

☐ Evaluate ROI every 30 days

☐ Update AI policy quarterly

☐ Train team continuously

☐ Document every automation

CHAPTER 16 ACTION PLAN

My first 90-day AI goal:

____

My first workflow to automate:

____

The tools I will use:

____

My end-of-year transformation goal:

____

____

Prompts Done For YOU!

This is a Special Added EXECUTION PLAN

Let's get Locked In!

- Every chapter gets a **dedicated "AI Prompts for Implementation" section**
- Prompts are:
    - Copy/paste ready
    - Tool-agnostic (ChatGPT, Claude, Gemini)
    - Written for **non-technical SMB owners**
    - Directly tied to the *action outcome* of that chapter
- These prompts will:
    - Turn the book into a **working playbook**
    - Dramatically increase perceived value
    - Become future lead magnets, bonuses, workshops, and courses

We start with **Chapter 4**, because this is where paralysis turns into momentum.

CHAPTER 4

AI PROMPTS FOR IMPLEMENTATION

The Overwhelm Problem — "I Don't Know Where to Start"

How to use these prompts:
Open ChatGPT (or your preferred AI tool).
Copy and paste **one prompt at a time.**
Replace the brackets with your business details.
Do not overthink it. Momentum beats perfection.

PROMPT 1 — Diagnose Your Biggest AI Opportunity

Purpose: Identify where AI will deliver the fastest ROI.

Prompt:

I run a **[type of business]** with **[number] employees.**
List the top **10 repetitive or frustrating tasks** that currently cost me the most **time, money, or customer goodwill** each week.
Rank them from **highest to lowest potential ROI if automated or assisted by AI.**

What you should get:
A ranked list of pain points instead of vague anxiety.

PROMPT 2 — Find the One Workflow to Start With

Purpose: Apply the *Single Workflow Rule*.

Prompt:

From the list above, help me choose **ONE workflow** that:

- happens at least weekly
- affects customers or revenue
- is easy to explain in words
- carries low risk if improved with AI
  Explain *why* this workflow is the best place to start.

What you should get:
Clarity. One target. No overwhelm.

PROMPT 3 — Define Success in 30 Days

Purpose: Create a measurable win.

Prompt:

If I successfully improved or automated this workflow using AI, define what **success should look like in 30 days**.
Include:

- time saved
- money saved or earned

- customer experience improvements
- employee workload changes
  Keep it realistic for a small business.

What you should get:
Clear metrics instead of vague hope.

PROMPT 4 — Choose the Right First AI Tool

Purpose: Prevent tool overload and wasted money.

Prompt:

Recommend **1–2 simple AI tools under $50/month** that could help me improve this workflow.
Explain:

- what each tool does
- why it's beginner-friendly
- what I should *not* expect it to do yet
  Assume I have no technical background.

What you should get:
A short, sane shortlist — not 20 options.

PROMPT 5 — Design a Safe First Automation

Purpose: Avoid breaking things.

Prompt:

Design a **very small, low-risk AI-assisted workflow** for this task.
The workflow should:

- keep a human in the loop
- avoid sensitive customer data
- be reversible if needed
- take less than one week to test
  Show the steps in plain English.

What you should get:
A pilot system, not a disaster.

PROMPT 6 — Assign the "AI Champion"

Purpose: Ensure implementation actually happens.

Prompt:

Based on a small business like mine, describe the ideal qualities of an **AI Champion** on my team.
Then suggest how I should train that person over the next **7 days** to own this workflow.

What you should get:
Ownership instead of abandonment.

PROMPT 7 — Momentum Check

Purpose: Lock in confidence and forward motion.

Prompt:

If I successfully complete this one AI workflow, explain:

- how it will change my confidence as an owner
- how it prepares me for the next AI implementation
- why starting small is actually a long-term advantage

What you should get:
Psychological momentum — which matters more than tech.

WHAT YOU NOW HAVE (AND MOST BOOKS NEVER GIVE)

You have:

- Identified your **first AI use case**
- Chosen **one workflow**
- Defined **clear ROI**
- Picked a **safe tool**
- Designed a **pilot system**
- Assigned ownership

- Eliminated overwhelm

This is where **skeptics are converted into practitioners**.

CHAPTER 5

AI PROMPTS FOR IMPLEMENTATION

Your First 30 Days With AI: A Simple, Stress-Free Blueprint

How to use these prompts:
Paste **one prompt at a time** into ChatGPT (or Claude / Gemini).
Do not combine prompts.
Answer honestly.
Speed beats perfection.

PROMPT 1 — Confirm Your First 30-Day Focus

Purpose: Prevent scope creep and burnout.

Prompt:

I am committing to a **30-day AI test**.
Based on my business (**[type of business]**), confirm whether my chosen workflow (**[describe it]**) is:

- realistic for 30 days
- measurable

- low risk
- likely to produce early ROI
  If not, suggest a *simpler* alternative.

Outcome:
Confidence that you're not trying to do too much.

PROMPT 2 — Break the Workflow Into AI-Friendly Tasks

Purpose: Turn "overwhelm" into steps AI can handle.

Prompt:

Break this workflow into **small tasks** and label each one as:

- Human-only
- AI-assisted
- Fully automated (later)
  Explain *why* for each task.

Outcome:
A clear division of labor between humans and AI.

PROMPT 3 — Build the Week-by-Week AI Plan

Purpose: Lock in momentum.

Prompt:

Create a **simple 4-week plan** for improving this workflow using AI.

- Week 1: setup & learning
- Week 2: first automation
- Week 3: testing & adjustment
- Week 4: measurement & decision
  Keep it practical and low stress.

Outcome:
A roadmap you can follow without guessing.

PROMPT 4 — Design the First Automation (Safely)

Purpose: Avoid breaking systems or confusing customers.

Prompt:

Design **one AI-assisted automation** for this workflow that:

- keeps a human review step
- does not touch sensitive data
- can be turned off easily
- improves speed or accuracy
  Describe it step-by-step in plain English.

Outcome:
A real system — not theory.

PROMPT 5 — Train the AI Champion

Purpose: Ensure adoption actually sticks.

Prompt:

I have chosen **[role/person]** as my AI Champion.
Create a **7-day learning plan** for them that includes:

- what they should learn
- what they should practice
- what success looks like
  Assume no technical background.

Outcome:
Ownership and accountability.

PROMPT 6 — Define ROI the Right Way

Purpose: Measure what matters.

Prompt:

Help me define **3–5 simple metrics** to track whether this AI workflow is working.
Include:

- time saved
- customer impact

- revenue or cost implications
  Avoid vanity metrics.

Outcome:
Proof instead of feelings.

PROMPT 7 — Decide What to Do After Day 30

Purpose: Prevent stalls after early success.

Prompt:

If this workflow succeeds after 30 days, recommend:

- whether I should scale it
- whether I should automate further
- what the *next* workflow should be
  Keep recommendations conservative and realistic.

Outcome:
A natural bridge to Chapter 6 (Tech Stack).

PROMPT 8 — Reset Expectations (Critical)

Purpose: Stop perfectionism from killing progress.

Prompt:

Remind me what a **successful first 30 days with AI** actually looks like for a small business — and what it does *not* look like.

Outcome:
Confidence instead of self-sabotage.

CHAPTER 6 — THE SMB AI TECH STACK

AI PROMPTS FOR IMPLEMENTATION

Purpose of This Chapter's Prompts

Help readers build a **lean, non-wasteful AI stack** without shiny-object syndrome.

Core Outcome

A **3–5 tool stack**, chosen intentionally, integrated correctly.

Prompts

Prompt 1 — Define My Non-Negotiable Needs

Based on my business type (**[type]**) and size (**[employees/revenue]**), list the **5 core problems** my tech stack must solve this year.

Prompt 2 — Eliminate Tool Overload

From the AI tool categories below, tell me which **I should ignore for now** and why.
Categories: marketing, sales, customer service, ops, HR, finance, analytics.

Prompt 3 — Build My Lean Stack

Recommend a **minimal AI tech stack (3–5 tools)** that solves my biggest problems without overlap or complexity.

Prompt 4 — Stack Integration Map

Show how these tools should connect at a **high level** so data flows cleanly.

Prompt 5 — Cost vs ROI Reality Check

Estimate realistic ROI for each tool within 90 days. Flag anything unlikely to pay for itself.

CHAPTER 7 — AI FOR MARKETING & LEAD GENERATION

AI PROMPTS FOR IMPLEMENTATION

Core Outcome

A **repeatable, automated marketing engine**.

Prompt 1 — Clarify My Ideal Customer

Help me define my **ideal customer profile** based on how people actually buy from my business.

Prompt 2 — Build My Brand Voice

Create a brand voice guide for my business that reflects trust, clarity, authority, and warmth.

Prompt 3 — Generate a 90-Day Content Plan

Create a 90-day content plan focused on leads, trust, and visibility — not vanity metrics.

Prompt 4 — Lead Capture Audit

Analyze where I may be losing leads due to slow response or poor follow-up.

Prompt 5 — Build an AI Follow-Up Sequence

Write a short email + SMS follow-up sequence that increases conversions without sounding robotic.

CHAPTER 8 — AI FOR SALES & FOLLOW-UP

AI PROMPTS FOR IMPLEMENTATION

Core Outcome

A **sales system that never forgets to follow up**.

Prompt 1 — Diagnose My Sales Leaks

Identify where leads likely fall through the cracks in my current sales process.

Prompt 2 — Build My Ideal Follow-Up Timeline

Create a simple follow-up timeline optimized for modern buyer behavior.

Prompt 3 — Write Human-Sounding Sales Messages

Write follow-up messages that feel helpful, not pushy.

Prompt 4 — Lead Scoring Simplified

Explain how I should prioritize leads using basic AI scoring logic.

Prompt 5 — Owner Time Protection

Show how AI can protect my time without damaging relationships.

CHAPTER 9 — AI FOR CUSTOMER SERVICE & APPOINTMENTS

AI PROMPTS FOR IMPLEMENTATION

Core Outcome

Fast, accurate, consistent customer experiences.

Prompt 1 — Customer Pain Audit

Identify the most frustrating moments in my customer experience.

Prompt 2 — Build My AI Front Desk

Design a basic AI receptionist workflow for my business.

Prompt 3 — FAQ Intelligence

Create a customer FAQ that reduces repetitive questions.

Prompt 4 — No-Show Reduction

Suggest AI-powered reminders that reduce missed appointments.

CHAPTER 10 — AI FOR OPERATIONS & ADMIN

AI PROMPTS FOR IMPLEMENTATION

Core Outcome

Less chaos. More control.

Prompt 1 — Identify Operational Bottlenecks

List the operational tasks causing the most friction.

Prompt 2 — Automate Documentation

Show how AI can simplify SOPs and internal documentation.

Prompt 3 — Admin Task Elimination

Identify admin tasks I should stop doing manually.

CHAPTER 11 — AI FOR HR & HIRING

AI PROMPTS FOR IMPLEMENTATION

Core Outcome

Better hires, less burnout.

Prompt 1 — Improve Job Descriptions

Rewrite my job descriptions to attract better candidates.

Prompt 2 — Resume Screening

Show how AI can safely help screen candidates.

Prompt 3 — Onboarding System

Build a simple onboarding workflow powered by AI.

CHAPTER 12 — AI FOR FINANCE & ACCOUNTING

AI PROMPTS FOR IMPLEMENTATION

Core Outcome

Visibility without obsession.

Prompt 1 — Cash Flow Clarity

Help me understand my cash flow risks.

Prompt 2 — Expense Categorization

Show how AI reduces accounting errors.

Prompt 3 — Financial Summaries

Create a weekly financial snapshot for decision-making.

CHAPTER 13 — THE AI-ENABLED WORKFORCE

AI PROMPTS FOR IMPLEMENTATION

Core Outcome

Empowered employees, not replaced ones.

Prompt 1 — AI Role Mapping

Identify which tasks AI should assist each role with.

Prompt 2 — Change Management Script

Help me explain AI adoption to my team in a positive way.

Prompt 3 — Training Roadmap

Create a simple AI upskilling plan for staff.

CHAPTER 14 — AI SAFETY, COMPLIANCE & TRUST

AI PROMPTS FOR IMPLEMENTATION

Core Outcome

Confidence and control.

Prompt 1 — AI Risk Audit

Identify where AI could pose risks in my business.

Prompt 2 — Guardrails

Help me design simple rules for safe AI use.

Prompt 3 — Customer Trust

Draft a transparent explanation of how my business uses AI responsibly.

CHAPTER 15 — CHOOSING THE RIGHT TOOLS (NO SCAMS)

AI PROMPTS FOR IMPLEMENTATION

Core Outcome

Smart buying decisions.

Prompt 1 — Vendor Vetting

Create a checklist to evaluate AI tools honestly.

Prompt 2 — Red Flag Detector

Identify signs a tool is overhyped or risky.

Prompt 3 — Buy vs Skip

Tell me which AI categories I should avoid right now.

CHAPTER 16 — THE 12-MONTH AI TRANSFORMATION PLAN

AI PROMPTS FOR IMPLEMENTATION

Core Outcome

A **clear, realistic roadmap.**

Prompt 1 — 12-Month Plan

Build a phased 12-month AI adoption plan for my business.

Prompt 2 — Milestones

Define milestones that indicate success.

Prompt 3 — Long-Term Advantage

Explain how consistent AI use compounds over time.

#	Prompt focus	Primary business benefit
1	Business diagnosis	Spot quick wins and bottlenecks.
2	Positioning	Sharper messaging and differentiation.

#	Prompt focus	Primary business benefit
3	Offer validation	Reduce risk before launching.
4	90-day plan	Clear execution roadmap.
5	Customer language	Higher-converting copy.
6	Content calendar	Consistent marketing output.
7	Social posts	Better reach and engagement.
8	Nurture emails	Turn leads into customers.
9	SEO article	Organic traffic and authority.
10	Repurposing	More content from same work
11	Cold email	Open doors with new prospects.
12	Lead follow-up	Fewer leads slipping through.
13	Objection scripts	Higher close rates.

#	Prompt focus	Primary business benefit
14	Landing page	Better conversion from traffic.
15	Call prep	Stronger sales conversations.
16	Support templates	Faster, more consistent service.
17	SOP drafting	Smoother, scalable operations.
18	Vendor comparison	Smarter purchasing decisions.
19	Hiring guide	Better hires, less guesswork.
20	Pricing strategy	Higher revenue and margins.
21	Financial model	Visibility into cash and profit.
22	AI ROI	Justify or kill tech projects.
23	Inbox triage	Time savings for the owner.

#	Prompt focus	Primary business benefit
24	Decision memo	Clearer thinking on big choices.
25	Weekly review	Continuous improvement rhythm.

Every small business owner should have a small arsenal of reusable AI prompts that cover strategy, marketing, sales, operations, finance, HR, and personal productivity. Below are 25 high-leverage "plug-and-play" prompts, each with fill-in blanks you can adapt directly in ChatGPT, Claude, Gemini, etc.

How to Use These Prompts

Use this structure with every prompt for better results:
"Act as [role]. Here is my business: [1–3 lines context]. Here is the task: [prompt]."

You can swap in your niche, numbers, and real docs (pasting or linking where safe) to make each one sharper.

1–4: Strategy & Positioning

1. Business diagnosis & quick wins
   "Act as a small-business strategy consultant. My business: [describe industry, size, offer, main

customer]. Our top 3 challenges: [challenge 1–3]. Analyze why these problems exist, list the 5 fastest low-cost fixes I can implement in the next 30 days, and rank them by impact vs. effort in a table."

2. Offer and positioning sharpening
"Act as a positioning expert. For my business [describe], define:

- A clear one-sentence value proposition
- 3 core customer personas with pains, desired outcomes, and buying triggers
- 3 concrete differentiators vs. typical competitors in [market].
Keep language simple enough for a non-technical owner."

3. New product or service validation
"Act as a market researcher. I'm considering launching [new offer] for [target audience] at [price]. List:

- 5 reasons they would buy
- 5 objections
- 5 ways to test demand in the next 2 weeks with minimal cost (emails, landing pages, offers, etc.)."

4. Quarterly growth plan
"Act as a growth advisor for a small business doing about [annual revenue]. In one page, draft a 90-day

growth plan focused on [goal: e.g., more leads / higher-margin sales / better retention]. Organize into weekly milestones, key metrics to track, and what to stop doing that's low impact."

5–10: Marketing & Content

5. Customer research & language mining
   "Act as a customer-research specialist. For my business [describe], generate:

- 10 likely customer pain statements in their exact words
- 10 desired outcome statements
- 10 search phrases they'd type into Google when looking for my solution.
  Use casual, conversational language."

6. Content strategy & calendar
   "Act as a content marketing strategist. Given my business [describe] and target audience [describe], create a 4-week content plan with:

- 2 blog topics per week
- 3 social posts per week (mix of educational, proof, and offers)
- 1 email per week.
  Put everything in a table with title, hook, call-to-action, and channel."

7. High-performing social posts
   "Act as a social media copywriter for [platform: Instagram, LinkedIn, etc.]. Create 10 posts for [target audience] about [topic or offer].
   Each post should include:

- A thumb-stopping first line
- 2–4 short sentences of value
- A simple call-to-action pointing to [website/offer].
  Keep the tone [choose: conversational/professional/funny]."

8. Email newsletter or nurture sequence
   "Act as an email marketer. My list: [who they are]. My main offer: [describe]. Create a 5-email nurture sequence to turn new subscribers into buyers. For each email, provide: subject line, 2–3 sentence summary, and clear CTA. Emphasize education and trust, not pushy sales."

9. SEO blog outline & draft
   "Act as an SEO content writer. Target keyword: '[keyword]'. Audience: [describe].
   Step 1: Suggest 5 angle ideas with different search intents.
   Step 2: For the best angle, create an outline with H2/H3s and bullet points.
   Step 3: Draft a 1,000-word article in a simple, skimmable style that a non-expert can understand."

10. Marketing asset repurposing
"Act as a content repurposing specialist. Here is a long-form asset: [paste blog, webinar notes, or transcript]. Turn it into:

- 5 LinkedIn posts
- 5 short social captions
- 3 email ideas.
Adapt language to fit each channel and avoid repetition."

11–15: Sales & Lead Generation

11. Cold email that gets replies
"Act as a B2B sales copywriter. Write 3 versions of a cold email to [role] at [type of company] about [your product/service]. Mention that they likely struggle with [key pain]. Keep it under 120 words, not salesy, and end with a single yes/no question that makes it easy to reply."

12. Inbound lead follow-up sequence
"Act as a sales strategist. A lead just [action: filled out form, downloaded guide, requested quote]. Create a 4-touch follow-up sequence (email or SMS) over 10 days. For each touch, provide timing, message objective, and sample copy under 100 words."

13. Objection handling playbook
"Act as a sales coach. My offer: [describe]. Common

objections: [price/time/trust/etc.]. For each objection, explain the real fear behind it, then give 2–3 short talk-tracks I can use in calls or emails that are empathetic, not pushy."

14. Sales page / landing page draft
    "Act as a direct-response copywriter. Write a landing page for [offer] targeted at [audience]. Include:

- Headline and subheadline
- Problem section
- Solution and benefits
- 3 short testimonials (invent realistic ones I can later replace)
- Clear pricing and call-to-action.
  Keep paragraphs short and scannable."

15. Sales call prep & questions
    "Act as a sales advisor. I'm about to have a call with a prospect: [describe who they are and what they asked about]. Generate:

- 10 discovery questions to uncover budget, authority, need, and timing
- 5 questions to understand their current solution and frustrations
- A 3-sentence way to pitch my offer tailored to their likely priorities."

16–19: Operations, Service & HR

16. Customer service templates & FAQ
"Act as a customer support lead. For my business [describe], list the 15 most likely customer questions and complaints. Then write email templates and short chat responses for each that are friendly, clear, and de-escalating, with a consistent tone of voice."

17. Process and SOP drafting
"Act as an operations consultant. I want to create a standard operating procedure for [process: onboarding a new client, shipping an order, closing the books, etc.]. Turn my rough notes [paste notes or describe steps] into a clear SOP with: purpose, owner, inputs, step-by-step checklist, and quality checks."

18. Vendor or proposal comparison
"Act as an analyst. I'm choosing between [2–3 vendors/tools]. Here are the proposals/details: [paste or summarize]. Create a comparison table with price, key features, risks, contract terms, and pros/cons by scenario (tight budget, fastest implementation, best long-term value). End with a recommendation and reasoning."

19. Job description & interview guide
"Act as an HR generalist for a small business. I need to hire a [role]. My company: [describe]. Responsibilities: [list]. Create:

- A clear, inclusive job description

- 8–10 behavioral interview questions
- 5 red flags to watch for in candidates for this role."

20–22: Finance, Pricing & ROI

20. Pricing strategy exploration
    "Act as a pricing strategist. My current offer is [describe] at [current price]. Customers are [segment] and competitors charge [range]. Propose 3 alternative pricing strategies (e.g., tiered, value-based, retainer, bundles) with pros/cons, who each is best for, and example price points."

21. Basic financial model & scenarios
    "Act as a small-business financial analyst. Build a simple 12-month projection for my business with assumptions I'll fill in: leads per month, close rate, average order value, cost of goods, fixed overhead. Show: revenue, gross profit, and net profit by month in a table, and create best-, base-, and worst-case scenarios."

22. AI or software ROI estimate
    "Act as an AI ROI analyst for a small business. I'm considering [tool or automation] to replace [manual process]. Current volume: [number of tasks/hours per week], current cost: [labor rate or hours]. Tool cost: [license/implementation]. Estimate payback period, 12-month ROI, and list assumptions so I can adjust the numbers."

23–25: Owner Productivity & Decision Support

23. Inbox triage & response drafting
    "Act as my executive assistant. Here are 10 recent emails I need to reply to today: [paste or summarize]. For each, suggest:

- Priority (high/medium/low)
- Whether to reply, delegate, or ignore
- A 3–5 sentence reply I can customize and send."

24. Decision memo for a tough choice
    "Act as a decision coach. I'm deciding between [option A] and [option B] about [issue]. Summarize the decision in a one-page memo with: context, options, pros/cons, risks, likely second-order effects, and a recommendation. Use bullet points and plain language suitable for a small-business owner."

25. Weekly review & planning
    "Act as an accountability coach for a small-business owner. I will paste my last week's activities, wins, and frustrations. Turn them into:

- 3 key lessons
- 3 things to stop doing
- 3 priorities for next week

A simple daily schedule block for deep work, admin, and sales/marketing."

Effective prompts are not about "speaking robot"; they are about asking better questions so AI can behave like a sharp assistant who already understands your business. With a few simple habits, even a busy, overwhelmed owner can turn AI from noise into a practical, everyday tool.

The Philosophy: Think "Briefing, Not Magic Spell"

AI works best when you treat it like a new hire you're briefing, not a genius that can read your mind. If you give sloppy instructions, you get sloppy output—no matter how advanced the model is.

For small business owners, that means:

- You do not need to learn "AI language"; you just need to be clear about what you want, why you want it, and who it is for.
- A good prompt is a conversation starter you refine, not a one-shot incantation you must perfect on the first try.

What Goes into a Valuable Prompt

Most strong prompts share a simple structure: Persona → Task → Context → Constraints → Format.

1. Persona (who the AI should act as)

    - Example: “Act as a marketing manager for a local service business...”

    - Why it matters: It instantly orients the AI toward the right tone, depth, and point of view.

2. Task (what you actually want done)

    - Be explicit: “Write,” “analyze,” “summarize,” “compare,” “draft a plan,” “turn this into an email,” etc.

    - Vague tasks like “help me with marketing” produce vague answers; precise tasks produce focused, usable outputs.

3. Context (the business and situation)

    - Briefly describe your business, audience, and goal: industry, offer, price range, type of customer, and current challenge.

    - Context is what turns generic advice into something that sounds like it knows your world.

4. Constraints (boundaries and requirements)

    - Examples: word count, tone (“plain English, no jargon”), region, budget limits, legal constraints, or time frame.

- Constraints keep the answer from wandering and make it directly usable.

5. Format (how you want the output delivered)

    - Ask for bullets, tables, step-by-step checklists, email drafts, scripts, or templates.
    - A clear format makes it easy to copy/paste into your systems without extra editing.

Principles That Make Prompts "Work" for Business

Think of these as guardrails so AI becomes a multiplier, not a distraction.

- Be specific, not short.
  One well-structured paragraph with persona, task, context, and constraints beats a three-word prompt every time.
- Break big problems into small prompts.
  Instead of "Create my whole marketing strategy," start with: "List my top 5 customer segments," then "Now draft one email for segment A," and so on.
- Iterate instead of starting over.
  Treat each response as version 1. Say: "Make this shorter," "Use simpler language," or "Rewrite for restaurant owners instead of consultants."

- Show, don't just tell.
  Paste real examples: a past email, a landing page, a review. Ask AI to "match this style but improve clarity" or "rewrite this for [new audience]."

- Align with business goals.
  Tie every prompt to a clear outcome: more leads, faster support replies, clearer SOPs, better financial visibility, etc.
  This keeps you out of "AI toy" mode and in "business return" mode.

Making It Feel Easy, Not Overwhelming

Owners often feel they're "behind" on AI; in reality, a few good prompts used repeatedly can create real leverage. Studies on prompt good-practice show that even basic clarity and structure can improve relevance and productivity by large margins for non-technical users.

To keep it simple:

- Reuse your best prompts as templates instead of reinventing them every time.

- Save 5–10 prompts that cover your core needs (emails, offers, SOPs, planning) and treat them as part of your standard toolkit, like your CRM or calendar.

Use this checklist every time you prompt if you want steady, business-grade results instead of random "sometimes great, sometimes useless" answers. You can print or save this as your standard prompt workflow for owners and staff.

Step-by-Step Prompt Checklist

1. Define the outcome in one sentence

- Ask yourself: "What do I actually need from AI right now?" (example: "a 5-email sequence," "a comparison table," "a step-by-step plan").
- Write that as a single clear outcome at the top of your prompt before anything else.

"I want a 3-step plan to follow up with cold leads from last quarter."

2. Choose the role for the AI

- Start with "Act as..." or "You are..." and pick a role that fits the task: marketing manager, sales coach, HR generalist, operations consultant, etc.
- This instantly shifts tone, level of detail, and style toward what you actually need.

"Act as a small-business email marketing specialist."

3. Add concise business context

- In 2–4 sentences, describe:

    - What your business does
    - Who your customers are
    - Price range or service level
    - Any current constraint (local only, B2B only, etc.)
- Avoid essays; you want "good brief," not a full history.

"We're a local HVAC company serving homeowners in [city], with average jobs between $300–$1,500, focused on fast, reliable repairs."

4. Specify the exact task

- Use clear verbs: write, list, analyze, compare, summarize, draft, rewrite, improve, structure, prioritize.
- If there are multiple tasks, number them.

"1) Draft 3 follow-up emails; 2) Make each under 120 words; 3) Include a soft call-to-action to schedule a visit."

5. Set constraints and boundaries

- Common constraints that improve consistency:
    - Length (word count, number of bullets)
    - Tone (plain English, professional, friendly, no jargon)
    - Audience (non-technical owner, first-time buyer, CFO, etc.)

    - Region or regulations if relevant (U.S., EU, specific state)
- Also say what you *don't* want (no emojis, no fluff, no made-up stats).

"Use simple, non-technical language, 6th–8th grade reading level, no hype, no emojis."

6. Request a clear output format

- Tell AI *how* to present the answer:
    - Bullet list
    - Numbered steps
    - Table (great for comparisons)
    - Script, email, ad, SOP template, etc.
- This step alone makes outputs much easier to plug into your systems.

"Present the emails as three separate blocks with subject line + body, ready to copy-paste."

7. Provide examples or source material (optional but powerful)

- Paste: a past email, landing page, review, or SOP and say "match this style but improve clarity" or "rewrite this for [new audience]."

- When you show AI what "good" looks like in your world, you get more consistent style and tone.

"Here is an email that worked well before; use the same tone but adapt it for lapsed customers."

8. Add "do/ don't" quality rules

- Simple guardrails save you editing time:
    - "Do: focus on benefits, include 1 call-to-action."
    - "Don't: invent statistics, don't reference internal systems."
- For decisions, you can ask the model to explain its reasoning briefly.

"Explain your reasoning in 3 bullets so I can sanity-check your suggestions."

9. Run, review, and refine (iteration loop)

- Treat the first answer as version 1, not final.
- Give short, direct feedback. "Shorter," "More formal," "Focus only on email #1," "Add an example for HVAC maintenance plans."

This "prompt → output → refine" loop is where consistency really comes from over time.

10. Save winning prompts as templates

- When a prompt gives you a great, repeatable result, save it in a doc, note app, or shared "prompt library" for your team.
- Keep variables (like [CITY], [OFFER], [AUDIENCE]) in brackets so anyone can plug in their info quickly.

Over time this becomes a small set of "house prompts" your business can rely on for consistent outputs across people and projects.

INDEX

AI Is Coming for Your Business

N

O

P

APPENDIX

APPENDIX A — The SMB AI Readiness Checklist

A simple tool to determine whether your business is ready to start using AI today.

This checklist removes guesswork and gives owners a clear "green light" or "yellow light" before implementing AI.

Section 1: Business Readiness

Check all that apply:

☐ We have at least **one repetitive task** we want to eliminate

☐ We have a clear understanding of where we lose time

☐ We have a clear understanding of where we lose money

☐ We can identify one workflow that frustrates customers

☐ We can dedicate **1 hour per week** to AI setup for the first month

☐ We have a team member who can act as our "AI Champion"

Section 2: Technology Readiness

☐ We have a business email domain

☐ We use at least one digital tool (email, CRM, scheduling, etc.)

☐ We have basic comfort with typing, copy/paste, and web navigation

☐ We can log into cloud-based tools

☐ We can connect apps through Zapier (or are willing to learn)

Section 3: Financial Readiness

☐ We have $20–$200/month available for AI tools

☐ We understand AI tools are cheaper than labor

☐ We are comfortable testing a tool for 30–60 days

(This is not $200/month per tool. This is your ENTIRE AI budget.)

Section 4: Cultural Readiness

☐ Our team understands AI will *support* their work, not replace them

☐ We are open to new systems and workflows

☐ We are willing to test, refine, and experiment

☐ We accept that early mistakes are normal

Your Results

14–18 checks:
You are fully ready. Begin implementing AI immediately.

9–13 checks:
You are ready but need clarity on your first workflow. Start with Chapter 5.

0–8 checks:
You need preparation first. Revisit Chapters 3–4 and Appendix B.

APPENDIX B — The "One Workflow" Prioritization Worksheet

Your step-by-step tool for identifying which AI workflow to automate first.

This worksheet removes overwhelm and helps owners get fast, measurable wins.

STEP 1 — Identify Your Time Wasters

Mark any that apply:

☐ **Customer responses**

☐ **Scheduling & rescheduling**

☐ **Billing & invoicing**

☐ **Social media content**

☐ **Email writing**

☐ **Quoting & proposals**

☐ **Admin tasks**

☐ **Inventory tracking**

☐ **Documentation**

☐ **Reporting**

☐ **Hiring tasks**

☐ **Reviewing analytics**

STEP 2 — Identify Your Money Leaks

Mark any that apply:

☐ **Missed leads**

☐ **Slow follow-up**

☐ **Poor appointment flow**

☐ **Errors that cost you money**

☐ **Delayed billing**

☐ **Marketing inconsistency**

☐ **High labor cost for repetitive work**

STEP 3 — Identify Customer Friction Points

Mark any that apply:

☐ **Slow replies**

☐ **Incorrect answers**

☐ **Confusing booking process**

☐ **Inconsistent information**

☐ **Unclear next steps**

☐ **Poor post-service communication**

STEP 4 — Your Automation Score

Each box = 1 point.

- **0–5 points:** Start with AI writing (emails, content, templates)
- **6–10 points:** Start with customer response AI
- **11–15 points:** Start with AI CRM & follow-up

- **16+ points:** You need multi-department AI support (start with operations + CRM)

STEP 5 — Declare Your First Workflow

"I will automate this ONE workflow first:

____

___________________________."

This becomes your Month 1 focus.

APPENDIX C — The AI Policy Template (Fill-In-The-Blanks)

Protect your business, your customers, and your team with a simple, customizable AI policy.

Purpose of This Policy

This policy defines how AI may be used responsibly inside our business to ensure:

- data safety
- customer privacy
- accuracy
- compliance
- ethical standards
- employee alignment

1. Approved AI Tools

Our business approves the use of:

____

____

____

Employees may NOT use unapproved AI tools.

2. Allowed Use Cases

Employees may use AI for:

- email drafts
- customer replies (review required)
- marketing content
- research
- internal documentation
- scheduling support
- SOP creation
- job posting drafts

3. Restricted Use Cases

AI may NOT be used for:

- sharing customer PII
- generating legal documents without review

- financial decisions without review
- medical advice
- compliance decisions
- sending unreviewed messages to customers
- storing sensitive data

4. Human Review Rule

Every AI output that goes to:

- customers
- vendors
- prospects
- employees

...must be reviewed by a human.

This includes:

- emails
- texts
- proposals
- ads
- website content
- contracts

5. Data Handling

Employees must:

- remove private information before prompting

- use company-approved accounts
- follow password and security protocols
- avoid sharing proprietary data

6. Continuous Improvement

Our team will:

- review AI performance quarterly
- update tools as needed
- provide ongoing training
- revise prompts and workflows
- maintain version control of SOPs

APPENDIX D — 20 High-Impact Prompts for Small Businesses

A curated starter library to multiply productivity across every department.

These prompts are structured, reliable, and can be pasted directly into your AI tool.

Marketing Prompts

1. "Create 30 days of social media content for a **[business type]** targeting **[ideal customer]**."
2. "Rewrite this text in my brand voice: **[paste text]**."

3. “Generate 10 email subject lines that increase open rates.”
4. “Create 5 promotional ideas for **[holiday/season]**.”
5. “Write 10 short-form video hooks for **[topic]**.”

Customer Service Prompts

6. “Draft responses to FAQs based on this information: **[paste info]**.”
7. “Create polite, professional replies for customer complaints.”
8. “Write a message to re-engage customers who haven’t booked in 90 days.”
9. “Generate clearer instructions for: **[service/task]**.”

Sales Prompts

10. “Write a follow-up sequence for leads who haven’t replied in 24/48/72 hours.”
11. “Create an objection-handling script for: **[product/service]**.”
12. “Write a persuasive proposal template.”

13. “Rewrite this value proposition to be clearer and more compelling.”

Operations Prompts

14. “Summarize this document into bullet points: **[paste]**.”
15. “Create an SOP for **[task]** using this information.”
16. “Draft onboarding instructions for a new employee in **[role]**.”
17. “Organize this list into categories: **[paste list]**.”

Finance Prompts

18. “Summarize last month’s financial data into a simple report.”
19. “Create a cash-flow forecast based on this data: **[paste numbers]**.”
20. “Generate invoice reminders in a friendly but firm tone.”

APPENDIX E — The AI Tool Buyer’s Guide

How to evaluate any AI tool in under 5 minutes.

Use this scorecard before purchasing ANY tool:

1. Purpose (Score 1–10)

Does this tool solve one of my top 3 business problems?

2. Learning Curve (Score 1–10)

Can a non-technical employee use it?

3. Integration (Score 1–10)

Does it connect to Zapier, CRM, email, calendar?

4. ROI Speed (Score 1–10)

Will it produce value within 30 days?

5. Safety (Score 1–10)

Does it protect customer data?

6. Vendor Trust (Score 1–10)

Does the company have real customers, documentation, and support?

7. Scalability (Score 1–10)

Will it still be useful a year from now?

TOTAL SCORE (Out of 70)

- **60–70:** Buy immediately
- **45–59:** Safe to test
- **30–44:** Only if needed
- **Below 30:** Walk away

APPENDIX F — SMB-Friendly AI Glossary

A Clear, Plain-English Guide to Every AI Term a Business Owner Needs to Know

This glossary avoids technical jargon and gives real-world business meanings, examples, and explanations.
Everything here is phrased so a non-technical reader can understand instantly.

Let's begin.

A

AGI (Artificial General Intelligence)

A hypothetical future AI that can think, reason, and learn like a human across all tasks.
Not available today. Not used in business.

AI (Artificial Intelligence)

Software that can understand information, analyze patterns, make decisions, and generate responses — without needing step-by-step instructions.
The core technology powering tools like ChatGPT.

AI Agent

A software assistant that not only thinks but **acts** on your behalf.
Example: An AI that books appointments, follows up with leads, updates calendars, and messages customers automatically.

AI Ethics

Guidelines that ensure AI is used responsibly, safely, and transparently.
Usually includes:

- privacy
- fairness
- accuracy
- human oversight
- transparency

AI Integration

Connecting AI tools to your existing systems (CRM, email, website, scheduling).
Often done through Zapier, Make, or native integrations.

AI Model

The "brain" that powers AI tools.
Examples: GPT-4, Claude, Gemini.
Different models have different strengths (writing, reasoning, research, speed).

AI Prompt

The text you type into an AI tool to tell it what you want.
The better the prompt, the better the result.

B

Bias (AI Bias)

When an AI system gives unfair or skewed results due to biased training data.
Example: AI recommending certain candidates over others unfairly.

Bot

Short for "robot," usually refers to chatbots.
AI bots can respond to customers automatically on your website or social media.

Business Logic

The rules your business follows — pricing, hours, booking policies, service areas.
AI often needs your business logic to give correct answers.

C

Chatbot

A digital assistant that can answer questions on your website or messages.
Modern AI chatbots are conversational and can handle booking, quoting, FAQs.

ChatGPT

One of the most widely used AI tools for writing, summarizing, planning, brainstorming, and creating content.

Claude

A powerful AI model known for excellent writing quality, reasoning, and handling long documents.

Computer Vision

AI that can "see" images or video and understand what's in them.
Example: AI that scans receipts, identifies products, or analyzes photos.

Content Generation

AI creating marketing content for you — posts, emails, ads, blogs, scripts, images.

Context Window

The amount of text an AI model can remember at once.
Higher context = better long-form reasoning.

D

Data Privacy

Protecting customer information when using AI tools. Requires removing sensitive details from prompts unless using secure business accounts.

Decision Engine

AI making judgment calls based on rules or patterns.
Example: An AI deciding which employee to assign a job to.

Deep Learning

A type of AI that uses multiple layers (like a brain) to learn patterns.
This is how modern AI models learn.

Digital Twin

An AI-powered replica of a system (or even a person's style or writing) that can act similarly.
Businesses will eventually have "digital twin" employees handling routine tasks.

E

Embedding

The mathematical representation of text so AI can understand meaning and search it efficiently.
Useful for knowledge bases and internal documentation.

Error Rate

How often AI gives incorrect outputs.
Lower error rate = more trustworthy.

Ethical AI

See AI Ethics.

F

Fine-Tuning

Training an AI model on your specific content (emails, SOPs, policies) so it behaves more like your business.

Many tools allow lightweight fine-tuning without technical expertise.

Foundation Model

The main large AI model that powers smaller tools and features (GPT-4, Claude Opus, Gemini Ultra).

G

Generative AI

AI that can *create* content — text, images, video, audio, code. This is the type of AI most SMBs use for content creation.

GPT

"Generative Pre-trained Transformer."
The architecture behind ChatGPT.
Meaning: A model trained on patterns of language to generate new text.

GPU (Graphics Processing Unit)

Special computer hardware used to run AI.
You don't need to worry about GPUs unless you're building AI yourself.

H

Hallucination (AI Hallucination)

When an AI confidently gives incorrect or fictional information. Prevented with human review and proper prompting.

Human-in-the-Loop (HITL)

AI + human oversight.
The human approves, edits, or corrects AI output before it is used.

This is the gold standard for SMBs.

I

Inference

The process of the AI model generating a response.
If the AI is "thinking," that's inference happening.

Integration

(See AI Integration)

Intelligent Automation

Automation combined with AI decision-making.
Example: AI determines which leads are urgent, then routes them.

J

Job Automation

AI replacing repetitive tasks once done by humans.
Not jobs — *tasks*.

K

Knowledge Base

A repository of your business info:

- FAQs
- SOPs
- pricing
- policies
- instructions
- internal documents

AI uses this to give accurate responses.

Keyword Extraction

AI identifying important terms for SEO, marketing, and research.

L

LLM (Large Language Model)

The engine that powers AI tools like ChatGPT and Claude.
It "understands" and generates human-like text.

Lead Scoring (AI Lead Scoring)

AI ranking leads based on likelihood to buy.
Saves sales teams time and focuses their efforts.

Low-Code / No-Code

Tools that do not require programming skills.
Most SMB AI tools fall into this category.

M

Machine Learning (ML)

Software that learns from data over time.
Used for predictions, patterns, and analytics.

Model Drift

When an AI model becomes less accurate over time due to outdated data.
Avoided with continuous updates.

N

Natural Language Processing (NLP)

AI understanding human language — text or speech.

Neural Network

The structure inside AI models that mimics brain neurons.
Not needed for SMBs to understand deeply — but it powers everything.

O

Optimization

Improving AI workflows to increase accuracy, efficiency, and output.
Common in marketing, sales, and operations.

OCR (Optical Character Recognition)

AI reading text from images or PDFs.
Useful for scanning receipts or paperwork.

P

Parameter

A variable inside an AI model that influences how it "thinks."
Large models have billions of parameters.

Predictive Analytics

AI forecasting trends like:

- sales
- demand
- cash flow
- customer churn

Prompt Engineering

The skill of writing effective prompts.
The better the prompt → the better the AI output.

Q

Query

A request you send to AI.
Same as a prompt, but often used with databases or search engines.

R

Reinforcement Learning

A training method where AI learns from feedback or scoring. ChatGPT uses RLHF (Reinforcement Learning from Human Feedback).

Retrieval-Augmented Generation (RAG)

AI pulls information from your documents before answering. This is how businesses avoid hallucinations.

S

Scalability

AI's ability to grow with your business without extra cost or staff.

Sentiment Analysis

AI detecting emotional tone in customer messages.
Useful for support and reputation management.

Structured Data

Well-organized information (spreadsheets, forms, databases).

T

Training Data

The information an AI model learned from.
This determines intelligence, accuracy, and style.

Token

A chunk of text AI reads or generates.
Roughly:
1 token ≈ ¾ of a word
Tokens affect cost and context size.

U

Unstructured Data

Information not neatly organized:
- emails
- notes
- PDFs
- chat logs

AI excels at processing this.

User Intent

What a person *really* wants when they ask a question.
AI is trained to infer intent for better responses.

V

Vector Database

A database used to store embeddings (AI-readable meaning).
Used for advanced customer service AI and internal knowledge tools.

Voice AI

AI that interacts with customers using natural speech.
Example: Air.ai answering phone calls.

W

Workflow Automation

Using AI + tools like Zapier to automate steps across systems.
Example:
lead submits form → AI writes reply → CRM tags lead → text sent → appointment scheduled.

X / Y / Z

Zero-Shot Learning

AI completing a task it was not explicitly trained on.
Example: AI translating a language it's never seen before.

Zero-Touch Automation

Automations that run with no human involvement at all. Advanced teams use this for finance, operations, and fulfillment.

RESOURCES & TOOLS SECTION

How to Use This Section

Tools change.
Principles don't.

These tools are examples and not requirements.

AI Writing & Knowledge

- ChatGPT
- Claude
- Gemini
- Perplexity

Use for: writing, analysis, planning, communication

AI Marketing & Content

- Canva AI
- Jasper
- Adobe Firefly
- Descript

Use for: branding, visuals, campaigns

AI Customer Service & Scheduling

- Tidio
- Intercom
- Smith.ai

- Calendly

Use for: lead capture, booking, follow-up

CRM & Sales Automation

- HubSpot
- GoHighLevel
- Close

Use for: pipelines, follow-up, automation

Workflow Automation

- Zapier
- Make

Use for: system-to-system connections

Finance & Admin

- QuickBooks AI
- Wave
- Xero

HR & Hiring

- BambooHR AI
- Rippling AI

Learning & Staying Current

- Harvard Business Review (AI coverage)
- McKinsey Global Institute
- MIT Sloan Management Review

Ready to Turn AI from a Threat into a Competitive Advantage?

If this book has made one thing clear, it's this: **AI is not optional anymore.**
It is already reshaping how businesses operate, compete, serve customers, and control costs.

The real question isn't *whether* AI will impact your business,
it's **whether you'll lead the change or be forced to react to it.**

That's where **Mike Schiano** and **SP&S Contact Services** come in. With more than **30 years of executive leadership experience**, Mike has helped organizations navigate disruptive technology shifts, improve performance, and achieve aggressive growth goals long before AI became today's business imperative. He understands both sides of the equation: **the technology *and* the people who must make it work.**

Why Business Owners Work with Mike Schiano

- He translates **AI hype into real-world strategy** that drives measurable ROI
- He helps leaders **reduce costs without destroying culture or customer experience**
- He identifies where AI creates advantage and where it creates hidden risk
- He brings deep expertise in **business operations, customer experience, and business transformation**

Most importantly, Mike doesn't sell tools.
He helps you **make smart decisions** about where AI belongs in *your* business and where it doesn't.

AI will completely change how your customers interact with you and how your teams deliver value.
Companies that act now will dominate their markets.
Companies that wait will struggle to catch up.

Let's Talk About Your Business

If you're ready to:

- Future-proof your organization
- Improve operational performance
- Leverage AI without losing control of your brand, workforce, or customer experience

Now is the time to start the conversation.

If you found value in this book and are interested in exploring ways in which we can help you or your business, please reach out to us via LinkedIn, phone, or email. We invite you to listen to our podcast, ***Mike Schiano In the Queue***, available on major podcast platforms, and to follow our blog posts at spscontact.com

Contact us to discuss your particular questions, issues, and needs:

Email: mike@spscontact.com
Phone/Text: 470-560-0920
LinkedIn: linkedin.com/in/mikeschiano/
Blog & Insights: spscontact.com

Podcast: Mike Schiano ***In the Queue*** (available wherever you listen and download your favorite Podcasts).

Instagram:@MikeaboutMoney

AI is moving fast.
The smartest business leaders are moving faster.

Let's make sure you're one of them.

Mike Schiano

www.ingramcontent.com/pod-product-compliance
Lightning Source LLC
LaVergne TN
LVHW052021200926
848481LV00005B/1551